SpringerBriefs in Mathematics

SpringerBriefs in Mathematics showcases expositions in all areas of mathematics and applied mathematics. Manuscripts presenting new results or a single new result in a classical field, new field, or an emerging topic, applications, or bridges between new results and already published works, are encouraged. The series is intended for mathematicians and applied mathematicians.

More information about this series at http://www.springer.com/series/10030

Vyacheslav V. Chistyakov

Metric Modular Spaces

Theory and Applications

 Springer

Vyacheslav V. Chistyakov
National Research University
 Higher School of Economics
Department of Informatics, Mathematics
 and Computer Science
Nizhny Novgorod, Russia

ISSN 2191-8198   ISSN 2191-8201   (electronic)
SpringerBriefs in Mathematics
ISBN 978-3-319-25281-0   ISBN 978-3-319-25283-4   (eBook)
DOI 10.1007/978-3-319-25283-4

Library of Congress Control Number: 2015956774

Mathematics Subject Classification (2010): 46A80, 54E35, 26A45, 47H30, 26E25, 34A12

Springer Cham Heidelberg New York Dordrecht London
© Springer International Publishing Switzerland 2015

Printed on acid-free paper

Springer International Publishing AG Switzerland is part of Springer Science+Business Media (www.
springer.com)

To my family with love:
  Zinaida and Vasiliĭ (parents)
  Svetlana (wife)
  Dar'ya and Vasilisa (daughters)

Preface

The theory of *metric spaces* was created by Fréchet [39] and Hausdorff [43] a century ago. In its basis is the notion of *distance* between any two points of a set. Usually (but not necessarily), the algebraic structure of the set does not play any role in the metric space analysis. If the set under consideration has a rich algebraic structure, e.g., it is a linear space, metrics, or distance functions, on the set can be defined by means of *norms*. In particular, the theory of *Banach spaces* [5] (i.e., complete normed linear spaces) is of fundamental importance in modern functional analysis. Despite the well-known Kuratowski's theorem [58] on the embedding of a metric space X into a Banach space (of bounded functions on X), the language of metric spaces is indispensable in expressing *nonlinear* properties of various phenomena and objects in metric spaces. Many results of the Banach space theory are extended to the *metric linear space* theory (Rolewicz [95]).

At the same time, a century ago, Lebesgue's theory [59] of measure and integral was developed, which is the theory of Banach space L_1 of summable functions equipped with the L_1-norm. Lebesgue's theory was extended by Riesz in his paper on L_p-spaces [94] ($1 < p < \infty$). In [87, 88], Orlicz defined and studied his famous normed *Orlicz spaces* L_φ of φ-summable functions, where φ is a convex (Orlicz) function on the reals $\mathbb{R}$. For nonconvex functions φ, F-normed L_φ-spaces were introduced by Masur and Orlicz [71] and Musielak and Orlicz [77] in the context of modular spaces. From different perspectives, the theory of Orlicz spaces is presented in many monographs, e.g., Adams [1], Krasnosel'skiĭ and Rutickiĭ [56], Lindenstrauss and Tzafriri [65], Maligranda [68], Musielak [75], and Rao and Ren [92, 93].

Modular spaces are extensions of Lebesgue, Riesz, and Orlicz spaces of integrable functions. A general theory of *modular linear spaces* was founded by Nakano in his two monographs [80, 81], where he developed a spectral theory in semi-ordered linear spaces (vector lattices) and established the integral representation for projections acting in his modular space; Nakano's modulars on real linear spaces are convex functionals. Nonconvex modulars and the corresponding modular linear spaces were constructed by Musielak and Orlicz [77] (we refer to Musielak [75] for a more comprehensive account). Orlicz spaces and modular linear spaces have already become classical tools in modern nonlinear functional analysis.

In spite of the significant generality of the modular spaces theory over linear spaces or spaces equipped with an additional algebraic structure, the notions of *modular* (which in particular extends the notion of norm) and the corresponding *modular linear space* are too restrictive. This is concerned, e.g., with the problems from the multivalued analysis such as the definition of metric functional spaces, characterization of set-valued superposition operators, (pointwise) selection principles, and existence of regular selections of multifunctions (Chistyakov [11–21]).

The purpose of this contribution is to develop a general theory of modulars on *arbitrary sets* and present a comprehensive background on metric and topological properties of the corresponding modular spaces. In our approach, a *modular w* on a set X is a parametrized family $w = \{w_\lambda\}_{\lambda > 0}$ of functions of two variables of the form $w_\lambda : X \times X \to [0, \infty]$ satisfying certain natural axioms in both nonconvex and convex situations. On the one hand, if the family w is independent of parameter λ, we get the notion of (extended) metric w on X. On the other hand, if X is a linear space and $\rho : X \to [0, \infty]$ is a classical modular on X (in Nakano's or Musielak-Orlicz's sense), then the family w corresponding to $w_\lambda(x, y) = \rho((x-y)/\lambda)$ $(x, y \in X)$ defines a modular on X in our sense. Thus, our theory of modular spaces is consistent with the theories of metric spaces and modular linear spaces.

Depending on the context, modulars w allow different interpretations. For instance, the quantity $w_\lambda(x, y)$ may be thought of as a (absolute value of nonlinear) *mean velocity* between points x and y in time $\lambda > 0$. It is known that in classical Newtonian mechanics, the deterministic principle says that the initial state of a mechanical system (the collections of its positions and velocities at a certain moment of time) determines uniquely the whole movement of the system. Accordingly, a modular w on X generates a distance function between any two points of X (actually, several distance functions can be defined on X). A subset of X, where the distance function assumes finite values (and so becomes a metric), is called a *modular space*. A natural (canonical) modular on a metric space (X, d) is given by $w_\lambda(x, y) = d(x, y)/\lambda$, which is the "real" mean velocity, and the induced distance function is $d(x, y)$ on the modular space X. In this way, we restore the original metric space by means of the canonical modular. More modulars can be considered on the metric space X, e.g., $w_\lambda(x, y) = d(x, y)/\lambda^p$ $(p \geq 0)$, or $w_\lambda(x, y) = \exp(d(x, y)/\lambda) - 1$, and they generate different distance functions on X.

Naturally, a modular space endowed with the generated metric is a *metric modular space* (hence the title of the book), and so, the standard metric space theory and its terminology apply to it. However, modulars are far from being metrics in general: they do not satisfy the usual triangle inequality. Having the ability to efficiently generate metrics, modulars might have been called *premetrics* or *prametrics*. However, the last two terms are already in use in topology (see Deza and Deza [36, p. 4]). So having in mind the connections with the modular linear theory, we adopt a more adequate term *metric modular* (for w as above) for its full name, or simply *modular* as its abbreviation (if no ambiguity with classical modulars arises).

Being dependent on the parameter λ, modulars give rise to a nonmetric, more weak convergence on a modular space, called the *modular convergence*. In the modular linear space theory, this notion was introduced by Musielak and Orlicz [77]. Correspondingly, the modular space is a topological space equipped with the *modular topology*, which, as a rule, is nonmetrizable. These notions are more subtle, and we postpone their definition and discussion until Chap. 4.

We apply the approach to the theory of modular spaces on arbitrary sets based on the author's papers [14, 16, 18–29], and [31]. We have added many new results and examples and made the exposition as self-contained as possible. The prerequisites for the reading of this book are some background knowledge of real and functional analysis, linear algebra, and rudiments of general topology. Thus the text, or much of it, is quite accessible to the university undergraduate students.

The plan of the exposition is as follows. The material is tacitly divided into two parts: Theory (Chaps. 1, 2, 3 and 4) and Applications (Chaps. 5–6). In Chap. 1, we define the notion of (metric) modulars, give their classification, obtain elementary properties, and present many examples useful in the sequel. In Chap. 2, we treat the metrizability of modular spaces: in contrast to the modular theory on linear spaces, where only few norms and F-norms are known, we define infinitely many metrics on our modular space and study their properties and relations between them. Further extensions of the notion of modular are given in Chap. 3, where we also study tools (transforms), by means of which new modulars can be produced. The most important are the right/left inverse modulars exhibiting the "duality" between modular spaces (Theorem 3.3.8). Chapter 4 is devoted to the classical topological aspects of the modular spaces, connected with the metric and modular convergences. In Chap. 5, a special $\mathbb{N}$-valued pseudomodular is introduced, whose induced modular spaces are the sets of all bounded and regulated mappings on an interval. This pseudomodular is crucial for obtaining a powerful pointwise selection principle, from which previously known pointwise selection principles follow, including Helly's theorem. The final Chap. 6 addresses some important classes of mappings of bounded generalized variation, which we interpret as the modular spaces for specifically constructed modulars. The results include the description of superposition operators acting in modular spaces, the existence of regular selections of set-valued mappings, the new interpretation of Lipschitzian and absolutely continuous mappings, and the existence of solutions to the Carathéodory-type ordinary differential equations in Banach spaces with the right-hand side from the Orlicz space.

Each chapter ends with Bibliographical Notes and Comments containing appropriate references, comments, and supplementary material.

Acknowledgements I am indebted to Panos M. Pardalos (University of Florida, USA) for inviting me to publish with Springer and for constant encouragement. I thank Razia Amzad (Springer) for her assistence and Patrick Muldowney (Londonderry, Northern Ireland) for useful suggestions. My special thanks go to Lech Maligranda (Luleå, Sweden) for sending me (around 2000) the two-volume collection of Orlicz's original papers [89], which has influenced my research in the theory of modular spaces a lot. My work on this contribution was partially supported by LATNA Laboratory, NRU HSE, RF government grant, ag. 11.G34.31.0057.

Nizhny Novgorod, Bogorodsk, Russia   Vyacheslav V. Chistyakov
June 2015

Contents

1 Classes of Modulars.. 1
  1.1 Modulars Versus Metrics.. 1
  1.2 The Classification of Modulars 4
  1.3 Examples of Modulars 7
    1.3.1 Separated Variables 8
    1.3.2 Families of Extended (Pseudo)metrics 9
    1.3.3 Classical Modulars on Real Linear Spaces 11
    1.3.4 φ-Generated Modulars 13
    1.3.5 Pseudomodulars on the Power Set 15
  1.4 Bibliographical Notes and Comments................................ 16

2 Metrics on Modular Spaces 19
  2.1 Modular Spaces.. 19
  2.2 The Basic Metric 20
  2.3 The Basic Metric in the Convex Case 27
  2.4 Modulars and Metrics on Sequence Spaces 31
  2.5 Intermediate Metrics 35
  2.6 Bibliographical Notes and Comments................................ 43

3 Modular Transforms 45
  3.1 Variants of Modular Axioms and Metrics 45
    3.1.1 φ-Convex Modulars 46
    3.1.2 Modulars Over a Convex Cone 47
    3.1.3 Complex Modulars 48
  3.2 Miscellaneous Transforms 49
  3.3 Right and Left Inverses.. 50
  3.4 Convex Right Inverses 60
  3.5 Bibliographical Notes and Comments................................ 63

4   Topologies on Modular Spaces 65
　　4.1　The Metric Convergence and Topology 65
　　　　4.1.1　The Metric Convergence.. 65
　　　　4.1.2　The Metric Topology.. 66
　　4.2　The Modular Convergence 69
　　4.3　The Modular Topology 73
　　4.4　Bibliographical Notes and Comments................................ 78

5   Bounded and Regulated Mappings 79
　　5.1　The $\mathbb{N}$-Valued Pseudomodular 79
　　5.2　The Pointwise Selection Principle 87
　　5.3　Bibliographical Notes and Comments................................ 91

6   Mappings of Bounded Generalized Variation 93
　　6.1　The Wiener-Young Variation... 93
　　6.2　Lipschitzian Operators 99
　　6.3　Superposition Operators... 100
　　6.4　Selections of Bounded Variation...................................... 103
　　6.5　Absolutely Continuous Mappings 107
　　6.6　The Riesz-Medvedev Generalized Variation 110
　　6.7　Bibliographical Notes and Comments................................ 121

Appendix 123
　　A.1　Superadditive, Subadditive, and Convex Functions.................. 123
　　A.2　The Hausdorff Distance 126
　　A.3　Metric Semigroups and Abstract Convex Cones..................... 127

References.. 129

Index... 135

Acronyms

$\mathbb{R} = (-\infty, \infty)$	The set of all real numbers; ∞ means $+\infty$
$[a, b]$	The closed interval of real numbers
(a, b)	The open interval of real numbers
$(0, \infty)$	The set of all positive real numbers; $\lambda > 0$ means $\lambda \in (0, \infty)$
$[0, \infty)$	The set of all nonnegative real numbers; $\lambda \geq 0$ means $\lambda \in [0, \infty)$
$[0, \infty]$	The set of extended nonnegative numbers $[0, \infty) \cup \{\infty\}$
$a = \infty$	Means $a > \lambda$ for all $\lambda > 0$; $\infty \cdot 0 = 0$, and $\infty \cdot \lambda = \infty$ if $\lambda > 0$
$\mathbb{N}, \mathbb{Z}, \mathbb{Q}, \mathbb{C}$	The sets of all natural, integer, rational, complex numbers
$\varnothing$	The empty set; $\inf \varnothing = \infty$, $\sup \varnothing = 0$
$\equiv$	Identical to; equality by the definition
$g \equiv c$	$g(\lambda) = c$ for all $\lambda > 0$ if $g : (0, \infty) \to [0, \infty]$ and $c \in [0, \infty]$
$\Leftrightarrow$, iff	If and only if
w, w	The generic notation for a (pseudo)modular
w_{+0}, w_{-0}	The right, left regularization of w
w^+, w^-	The right, left inverse of w
x°	the center of modular spaces X_w^*, X_w^0, and X_w^{fin}

Spaces of mappings from $I = [a, b]$ into a metric space M (in decreasing order):

M^I	The set of all mappings $x : I \to M$
$B(I; M)$	Bounded mappings
$\mathrm{Reg}(I; M)$	Regulated mappings
$\mathrm{BV}_\varphi(I; M)$	Mappings of bounded Wiener-Young variation
$\mathrm{BV}(I; M)$	Mappings of bounded Jordan variation
$\mathrm{AC}(I; M)$	Absolutely continuous mappings
$\mathrm{GV}_\Phi(I; M)$	Mappings of bounded Riesz-Medvedev variation
$\mathrm{Lip}(I; M)$	Lipschitzian mappings

Spaces of mappings from $I = [a, b]$ into a Banach space M:

$L_1(I; M)$	Strongly measurable and Bochner integrable mappings
$L_\Phi(I; M)$	The Orlicz space of Φ-integrable mappings

Chapter 1
Classes of Modulars

1.1 Modulars Versus Metrics

In order to motivate the notion of *modular* on a set, we begin by recalling the notion of *metric*. Let X be a nonempty set.

A function $d : X \times X \to \mathbb{R}$ is said to be a *metric* on X if, for all elements (conventionally called points) $x, y, z \in X$, it satisfies the following three conditions (axioms):

(d.1) $x = y$ if and only if $d(x, y) = 0$ (nondegeneracy);
(d.2) $d(x, y) = d(y, x)$ (symmetry);
(d.3) $d(x, y) \leq d(x, z) + d(z, y)$ (triangle inequality).

The pair (X, d) is called a *metric space*. (Actually, two axioms suffice to define a metric, because conditions (d.1)–(d.3) are equivalent to (d.1), and (d.3) written in the 'strong' form as $d(x, y) \leq d(x, z) + d(y, z)$: in fact, putting $z = x$ and then interchanging x and y, we obtain (d.2). Traditionally, the symmetry property of d is introduced explicitly.)

Clearly, a metric d assumes nonnegative (and finite) values, and if X has at least two elements (which is tacitly assumed throughout), then $d \not\equiv 0$ on $X \times X$. If the value ∞ is allowed for d satisfying (d.1)–(d.3), then d is called an *extended metric* on X, and the pair (X, d) is called an *extended metric space*.

If $d : X \times X \to [0, \infty]$ satisfies (d.2), (d.3) and (only) a weaker condition

(d.1′) $d(x, x) = 0$ for all $x \in X$,

then d is called a *pseudometric* on X, and it is called an *extended pseudometric* on X if the value 'infinity' is allowed for d.

© Springer International Publishing Switzerland 2015
V.V. Chistyakov, *Metric Modular Spaces*, SpringerBriefs in Mathematics,
DOI 10.1007/978-3-319-25283-4_1

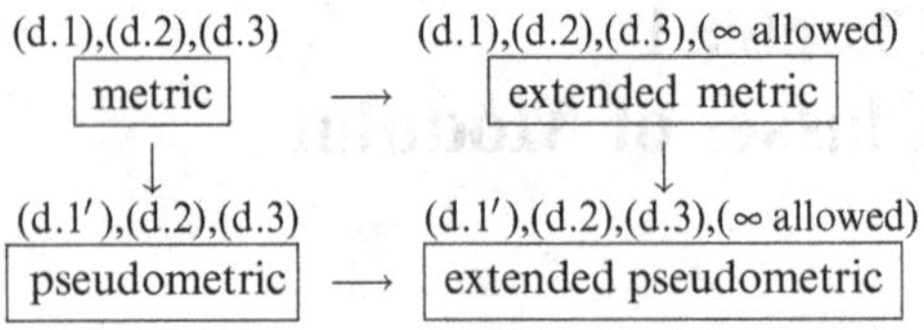

Fig. 1.1 Variants of metric
notions

(d.1),(d.2),(d.3)   (d.1),(d.2),(d.3),(∞ allowed)

| metric | $\longrightarrow$ | extended metric |

$\downarrow$ $\downarrow$

(d.1′),(d.2),(d.3)   (d.1′),(d.2),(d.3),(∞ allowed)

| pseudometric | $\longrightarrow$ | extended pseudometric |

The notion of metric reflects our *geometric* intuition of what a distance function on a set should be: to any two points $x, y \in X$, a number

$$0 \leq d(x, y) < \infty \qquad \text{(the } \textit{distance} \text{ between } x \text{ and } y)$$

is assigned, satisfying properties (d.1)–(d.3). The implications between the above four metric notions are presented in Fig. 1.1.

The idea of modular w on X can be expressed in *physical* terms as follows: to any parameter $\lambda > 0$, interpreted as time, and any two points $x, y \in X$, a quantity

$$0 \leq w_\lambda(x, y) \leq \infty \quad \text{(the } \textit{velocity} \text{ between } x \text{ and } y \text{ in time } \lambda)$$

is assigned, satisfying three axioms to be discussed below, and the one-parameter family $w = \{w_\lambda : \lambda > 0\} \equiv \{w_\lambda\}_{\lambda>0}$ of functions of the form $w_\lambda : X \times X \to [0, \infty]$ is a (generalized, nonlinear) *velocity field* on X.

Now we address the axioms of a modular. By a *scaling* of time $\lambda > 0$ we mean any value $h(\lambda) > 0$ such that the function $\lambda \mapsto \lambda/h(\lambda)$ is nonincreasing in λ (e.g., given $p \geq 1$, $h(\lambda) = \lambda^p$, or $h(\lambda) = \exp(\lambda^p) - 1$, or $h(\lambda) = \lambda e^\lambda$, etc.). Let (X, d) be a metric space, and $x, y \in X$. Consider the quantity

$$w_\lambda(x, y) = \frac{d(x, y)}{h(\lambda)} = \frac{\lambda}{h(\lambda)} \cdot \frac{d(x, y)}{\lambda} , \qquad (1.1.1)$$

which is a *scaled mean velocity* between x and y in time λ. For $h(\lambda) = \lambda$, this is the *mean* (or *uniform*) velocity, and so, in order to cover the distance $d(x, y)$, it takes time λ to move between x and y with velocity $w_\lambda(x, y) = d(x, y)/\lambda$.

The following natural-looking properties of quantity (1.1.1) hold.

(i) Two points x and y from X coincide (and $d(x, y) = 0$) if and only if any time $\lambda > 0$ will do in order to move from x to y with velocity $w_\lambda(x, y) = 0$ (that is, no movement is needed at any time). Formally, given $x, y \in X$, we have:

$$x = y \text{ if and only if } w_\lambda(x, y) = 0 \text{ for all } \lambda > 0 \text{ (nondegeneracy).}$$

(ii) For any time $\lambda > 0$, the mean velocity during the movement from point x to point y is equal to the mean velocity in the opposite direction, i.e., given $x, y \in X$,

$$w_\lambda(x, y) = w_\lambda(y, x) \text{ for all } \lambda > 0 \text{ (symmetry).}$$

(iii) The third property of quantity (1.1.1), which is, in a sense, a counterpart of the triangle inequality (for velocities!), is new and the most important. Suppose that movements from x to y happen to be made in two ways, but the *duration of time is the same* in each of the cases: (a) passing through a third point $z \in X$, or (b) moving directly from x to y. If λ is the time needed to move from x to z and μ is the time needed to move from z to y, then the corresponding mean velocities are equal to $w_\lambda(x, z)$ and $w_\mu(z, y)$. The total time spent during the movement in case (a) is equal to $\lambda + \mu$. It follows that the mean velocity in case (b) should be equal to $w_{\lambda+\mu}(x, y)$. It may become clear from the physical intuition that the velocity $w_{\lambda+\mu}(x, y)$ does not exceed at least one of the velocities $w_\lambda(x, z)$ or $w_\mu(z, y)$. This is expressed as

$$w_{\lambda+\mu}(x, y) \leq \max\{w_\lambda(x, z), w_\mu(z, y)\} \leq w_\lambda(x, z) + w_\mu(z, y) \qquad (1.1.2)$$

for all points $x, y, z \in X$ and times $\lambda, \mu > 0$. These inequalities can be verified rigorously: since $(\lambda + \mu)/h(\lambda + \mu) \leq \lambda/h(\lambda)$, it follows from (1.1.1) and (d.3) that

$$w_{\lambda+\mu}(x, y) = \frac{d(x, y)}{h(\lambda + \mu)} \leq \frac{d(x, z) + d(z, y)}{h(\lambda + \mu)} \leq \frac{\lambda}{\lambda + \mu} \cdot \frac{d(x, z)}{h(\lambda)} + \frac{\mu}{\lambda + \mu} \cdot \frac{d(z, y)}{h(\mu)}$$

$$= \frac{\lambda}{\lambda + \mu} w_\lambda(x, z) + \frac{\mu}{\lambda + \mu} w_\mu(z, y) \leq w_\lambda(x, z) + w_\mu(z, y). \qquad (1.1.3)$$

By (in)equality (1.1.3), conditions $w_\lambda(x, z) < w_{\lambda+\mu}(x, y)$ and $w_\mu(z, y) < w_{\lambda+\mu}(x, y)$ cannot hold simultaneously, which proves the left-hand side inequality in (1.1.2).

A *modular* on a set X is any one-parameter family $w = \{w_\lambda\}_{\lambda>0}$ of functions w_λ mapping $X \times X$ into $[0, \infty]$ and satisfying properties (i), (ii), and (iii) meaning (1.1.2). (The interpretation of modular as a generalized nonlinear mean velocity field has been chosen as the most intuitive and accessible; there are different interpretations of modular such as a double joint generalized variation of two mappings x and y.)

Even on a metric space (X, d), modulars may look unusual: given $\lambda > 0$ and $x, y \in X$, set $w_\lambda(x, y) = \infty$ if $\lambda \leq d(x, y)$, and $w_\lambda(x, y) = 0$ if $\lambda > d(x, y)$.

The difference between a metric (= distance function) and a modular (= velocity field) on a set is now clearly seen: a modular depends on a positive parameter λ and may assume infinite values (to say nothing of the axioms). The equality $w_\lambda(x, y) = \infty$ may be thought of as there is no possibility (or there is a prohibition) to move from x to y in time λ. For instance, the distance $d(x, y) = 10,000$ km between two cities x and y cannot be covered physically in $\lambda = 1, 2, \ldots, 100$ s; however, for times λ large enough, a certain finite velocity will do.

The essential property of a modular w (e.g., (1.1.1)) is that the velocity $w_\lambda(x, y)$ is *nonincreasing* as a function of time $\lambda > 0$.

A modular w on X gives rise to a *modular space* around a (chosen) point $x° \in X$—this is the set

$$X_w^* = \{x \in X : w_\lambda(x, x°) \text{ is finite for some } \lambda = \lambda(x) > 0\}$$

of those points x, which are *reachable* from $x°$ with a finite velocity. The knowledge of (mean) velocities $w_\lambda(x, y)$ for all $\lambda > 0$ and $x, y \in X$ provides more information than simply the knowledge of distances $d(x, y)$ between points x and y. In fact, if w satisfies (i), (ii) and the left-hand side (in)equality in (1.1.3), then the modular space X_w^* is metrizable by the following (implicit, or limit case) metric:

$$d_w^*(x, y) = \inf\{\lambda > 0 : w_\lambda(x, y) \leq 1\}.$$

Naturally, the pair (X_w^*, d_w^*) is called a *metric modular space*. For instance, the original metric space (X, d) is restored via the (mean velocity) modular (1.1.1) with $h(\lambda) = \lambda$ as the 'limit case' in that $X_w^* = X$ and $d_w^*(x, y) = d(x, y)$ for all $x, y \in X$.

This book is intended as the general study of modulars, modular spaces and metric modular spaces generated by modulars. Since the metric space theory is a well-established and rich theory, the main emphasis of this exposition is focused (where it is possible) on non-metric features of modulars and modular spaces.

Essentially, modulars serve two important purposes:

– to define new metric spaces, such as (X_w^*, d_w^*) and others, in a unified and general manner, and
– to present a new type of convergence in the modular space X_w^*, the so called *modular convergence*, whose topology is weaker (coarser) than the d_w^*-metric topology and, in general, is non-metrizable.

1.2   The Classification of Modulars

In the sequel, we study functions w of the form $w : (0, \infty) \times X \times X \to [0, \infty]$, where X is a fixed nonempty set (with at least two elements). Due to the disparity of the arguments, we may (and will) write $w_\lambda(x, y) = w(\lambda, x, y)$ for all $\lambda > 0$ and $x, y \in X$. In this way, $w = \{w_\lambda\}_{\lambda > 0}$ is a one-parameter family of functions $w_\lambda : X \times X \to [0, \infty]$. On the other hand, given $x, y \in X$, we may set $w^{x,y}(\lambda) = w(\lambda, x, y)$ for all $\lambda > 0$, so that $w^{x,y} : (0, \infty) \to [0, \infty]$. In the latter case, the usual terminology of Real Analysis can be applied to $w = \{w^{x,y}\}_{x,y \in X}$. For instance, the function w is called nonincreasing (right/left continuous, etc.) on $(0, \infty)$ if the function $w^{x,y}$ is such for all $x, y \in X$.

Definition 1.2.1. A function $w : (0, \infty) \times X \times X \to [0, \infty]$ is said to be a *metric modular* (or simply *modular*) on X if it satisfies the following three axioms:

(i) given $x, y \in X$, $x = y$ if and only if $w_\lambda(x, y) = 0$ for all $\lambda > 0$;
(ii) $w_\lambda(x, y) = w_\lambda(y, x)$ for all $\lambda > 0$ and $x, y \in X$;
(iii) $w_{\lambda+\mu}(x, y) \leq w_\lambda(x, z) + w_\mu(z, y)$ for all $\lambda, \mu > 0$ and $x, y, z \in X$.

Weaker and stronger versions of conditions (i) and (iii) will be of importance. If, instead of (i), the function w satisfies (only) a weaker condition

(i′) $w_\lambda(x, x) = 0$ for all $\lambda > 0$ and $x \in X$,

then w is said to be a *pseudomodular* on X. Furthermore, if, instead of (i), the function w satisfies (i′) and a stronger condition

(i_s) given $x, y \in X$ with $x \neq y$, $w_\lambda(x, y) \neq 0$ for all $\lambda > 0$,

then w is called a *strict modular* on X.

A modular (or pseudomodular, or strict modular) w on X is said to be *convex* if, instead of (iii), it satisfied the (stronger) inequality (iv):

(iv) $w_{\lambda+\mu}(x, y) \leq \dfrac{\lambda}{\lambda + \mu} w_\lambda(x, z) + \dfrac{\mu}{\lambda + \mu} w_\mu(z, y)$ for all $\lambda, \mu > 0$ and $x, y, z \in X$.

A few remarks concerning this definition are in order.

Remark 1.2.2. (a) The assumption $w : (0, \infty) \times X \times X \to (-\infty, \infty]$ in the definition of a pseudomodular does not lead to a greater generality: in fact, setting $y = x$ and $\mu = \lambda > 0$ in (iii) and taking into account (i′) and (ii), we find

$$0 = w_{2\lambda}(x, x) \leq w_\lambda(x, z) + w_\lambda(z, x) = 2w_\lambda(x, z),$$

and so, $w_\lambda(x, z) \geq 0$ or $w_\lambda(x, z) = \infty$ for all $\lambda > 0$ and $x, z \in X$.

(b) If $w_\lambda(x, y) = w_\lambda$ is independent of $x, y \in X$, then, by (i′), $w \equiv 0$. Note that $w \equiv 0$ is only a pseudomodular on X (by virtue of (i)).

If $w_\lambda(x, y) = w(x, y)$ does not depend on $\lambda > 0$, then axioms (i)–(iii) mean that w is an extended metric (extended pseudometric if (i) is replaced by (i′)) on X; w is a metric on X if, in addition, it assumes finite values.

(c) Axiom (i) can be written as $(x = y) \Leftrightarrow (w^{x,y} \equiv 0)$, and part ($i_\Leftarrow$) in it—as $(x \neq y) \Rightarrow (w^{x,y} \not\equiv 0)$. Condition ($i_s$) says that $(x \neq y) \Rightarrow (w^{x,y}(\lambda) \neq 0$ for all $\lambda > 0$), and so, it implies ($i_\Leftarrow$). In other words, (i_s) means that if $w_\lambda(x, y) = 0$ for some $\lambda > 0$ (and not necessarily for all $\lambda > 0$ as in ($i_\Leftarrow$)), then $x = y$. Thus, (i′) + (i_s) $\Rightarrow$ (i) $\Rightarrow$ (i′). Clearly, (iv) $\Rightarrow$ (iii). Thus, a (convex) strict modular on X is a (convex) modular on X, and so, it is a (convex) pseudomodular on X. These implications are shown in Fig. 1.2, and it will be seen later that none of them can be reversed.

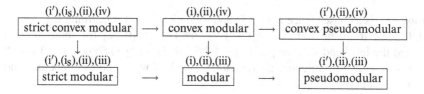

Fig. 1.2 Classification of modulars

(d) Rewriting (iv) in the form $(\lambda + \mu)w_{\lambda+\mu}(x, y) \leq \lambda w_\lambda(x, z) + \mu w_\mu(z, y)$, we see that the function w is a *convex* (pseudo)modular on X if and only if the function $\hat{w}_\lambda(x, y) = \lambda w_\lambda(x, y)$ is simply a (pseudo)modular on X. This somewhat unusual observation on the convexity of w will be justified later (see Sect. 1.3.3).

The *essential property* of a pseudomodular w on X is its monotonicity: given $x, y \in X$, the function $w^{x,y} : (0, \infty) \to [0, \infty]$ is *nonincreasing* on $(0, \infty)$. In fact, if $0 < \mu < \lambda$, then axioms (iii) (with $z = x$) and (i′) imply

$$w_\lambda(x, y) = w_{(\lambda-\mu)+\mu}(x, y) \leq w_{\lambda-\mu}(x, x) + w_\mu(x, y) = w_\mu(x, y). \tag{1.2.1}$$

As a consequence, given $x, y \in X$, at each point $\lambda > 0$ the limit from the right

$$(w_{+0})_\lambda(x, y) \equiv w_{\lambda+0}(x, y) = \lim_{\mu \to \lambda+0} w_\mu(x, y) = \sup\{w_\mu(x, y) : \mu > \lambda\} \tag{1.2.2}$$

and the limit from the left

$$(w_{-0})_\lambda(x, y) \equiv w_{\lambda-0}(x, y) = \lim_{\mu \to \lambda-0} w_\mu(x, y) = \inf\{w_\mu(x, y) : 0 < \mu < \lambda\} \tag{1.2.3}$$

exist in $[0, \infty]$, and the following inequalities hold, for all $0 < \mu < \lambda$:

$$w_{\lambda+0}(x, y) \leq w_\lambda(x, y) \leq w_{\lambda-0}(x, y) \leq w_{\mu+0}(x, y) \leq w_\mu(x, y) \leq w_{\mu-0}(x, y). \tag{1.2.4}$$

To see this, by the monotonicity of w, for any $0 < \mu < \mu_1 < \lambda_1 < \lambda$, we have:

$$w_\lambda(x, y) \leq w_{\lambda_1}(x, y) \leq w_{\mu_1}(x, y) \leq w_\mu(x, y),$$

and it remains to pass to the limits as $\lambda_1 \to \lambda - 0$ and $\mu_1 \to \mu + 0$.

Proposition 1.2.3. *Let w be a convex pseudomodular on X, and $x, y \in X$. We have:*

(a) *functions $\lambda \mapsto w_\lambda(x, y)$ and $\lambda \mapsto \lambda w_\lambda(x, y)$ are nonincreasing on $(0, \infty)$, and*

$$w_\lambda(x, y) \leq (\mu/\lambda)w_\mu(x, y) \leq w_\mu(x, y) \quad \text{for all} \quad 0 < \mu \leq \lambda; \tag{1.2.5}$$

(b) *if $w^{x,y} \not\equiv 0$ (e.g., w is a convex modular and $x \neq y$), then $\lim_{\mu \to +0} w_\mu(x, y) = \infty$;*
(c) *if $w^{x,y} \not\equiv \infty$, then $\lim_{\lambda \to \infty} w_\lambda(x, y) = 0$.*

Proof. (a) is a consequence of (1.2.1) and Remark 1.2.2(d) concerning $\hat{w}$.
(b) follows from the fact that $w_\lambda(x, y) = w^{x,y}(\lambda) \in (0, \infty]$ for some $\lambda > 0$ and the left-hand side inequality in (1.2.5): $w_\mu(x, y) \geq (1/\mu)\lambda w_\lambda(x, y)$ for all $0 < \mu < \lambda$. In particular, by Remark 1.2.2(c), if w is a convex modular and $x \neq y$, then $w^{x,y} \not\equiv 0$.
(c) Since $w_\mu(x, y) < \infty$ for some $\mu > 0$, the assertion is a consequence of the left-hand side inequality in (1.2.5). $\qquad\square$

Definition 1.2.4. Functions $w_{+0}, w_{-0} : (0, \infty) \times X \times X \to [0, \infty]$, defined in (1.2.2) and (1.2.3), are called the *right* and *left regularizations* of w, respectively.

Proposition 1.2.5. *Let w be a pseudomodular on X, possibly having additional properties shown in Fig. 1.2 on p. 5. Then w_{+0} and w_{-0} are also pseudomodulars on X having the same additional properties as w. Moreover, w_{+0} is continuous from the right and w_{-0} is continuous from the left on $(0, \infty)$.*

Proof. Since properties (i′), (ii), (iii), and (iv) are clear for w_{+0} and w_{-0}, we verify only (i$_\Leftarrow$), the strictness, and one-sided continuities.

Suppose w is a modular. Let $x, y \in X$, and $(w_{+0})_\mu(x, y) = 0$ for all $\mu > 0$. Given $\lambda > 0$, choose μ such that $0 < \mu < \lambda$. Then (1.2.4) and (1.2.2) yield $0 \le w_\lambda(x, y) \le w_{\mu+0}(x, y) = 0$, and so, by axiom (i), $x = y$. Now, assume that $(w_{-0})_\lambda(x, y) = 0$ for all $\lambda > 0$. Since $w_\lambda(x, y) \le w_{\lambda-0}(x, y) = 0$, axiom (i) implies $x = y$.

Let w be strict, and $x, y \in X$, $x \ne y$. By condition (i$_s$) and (1.2.4),

$$0 \ne w_\lambda(x, y) \le w_{\lambda-0}(x, y) \le w_{\mu+0}(x, y), \quad 0 < \mu < \lambda,$$

and so, $(w_{-0})_\lambda(x, y) \ne 0$ for all $\lambda > 0$ and $(w_{+0})_\mu(x, y) \ne 0$ for all $\mu > 0$.

Let us show that w_{+0} is continuous from the right on $(0, \infty)$ (the left continuity of w_{-0} is treated similarly). Since w_{+0} is a pseudomodular on X, it is nonincreasing on $(0, \infty)$, and so, if $\mu > 0$, $x, y \in X$, and $\gamma = (w_{+0})_{\mu+0}(x, y)$, we have, by (1.2.4), $\gamma \le (w_{+0})_\mu(x, y)$. In order to obtain the reverse inequality, we may assume that γ is finite. For any $\varepsilon > 0$ there exists $\mu_0 = \mu_0(\varepsilon) > \mu$ such that, if $\mu < \mu' \le \mu_0$, we have $(w_{+0})_{\mu'}(x, y) < \gamma + \varepsilon$. Given λ with $\mu < \lambda \le \mu_0$, choosing μ' such that $\mu < \mu' < \lambda$, we find, by virtue of (1.2.4), that $w_\lambda(x, y) \le w_{\mu'+0}(x, y) < \gamma + \varepsilon$. Passing to the limit as $\lambda \to \mu + 0$, we get the inequality $(w_{+0})_\mu(x, y) \le \gamma + \varepsilon$ for all $\varepsilon > 0$. □

Remark 1.2.6. In the above proof, we have shown that $(w_{+0})_{+0} = w_{+0}$ (as well as $(w_{-0})_{-0} = w_{-0}$), and one can show that $(w_{-0})_{+0} = w_{+0}$ and $(w_{+0})_{-0} = w_{-0}$.

1.3 Examples of Modulars

In order to get a better feeling of the notion of modular, we ought to have a sufficiently large reservoir of them. This section serves this purpose (to begin with). Where a metric notion is needed, we prefer a metric space context; generalizations to extended metrics and pseudometrics can then be readily obtained in a parallel manner. Instead of referring to the family $w = \{w_\lambda\}_{\lambda>0}$, it is often convenient and nonambiguous to term a (pseudo)modular the value $w_\lambda(x, y)$.

1.3.1   Separated Variables

Let (X, d) be a metric space, and $g : (0, \infty) \to [0, \infty]$ be an extended (nonnegative) valued function. We set

$$w_\lambda(x, y) = g(\lambda) \cdot d(x, y), \quad \lambda > 0, \quad x, y \in X, \tag{1.3.1}$$

with the convention that $\infty \cdot 0 = 0$, and $\infty \cdot a = \infty$ for all $a > 0$.

By (d.1) and (d.2), the family $w = \{w_\lambda\}_{\lambda>0}$ satisfies axioms (i$'$) and (ii). It follows that the modular classes of w on X (cf. Fig. 1.2) are characterized as follows:

(A) w is a pseudomodular on X iff axiom (iii) is satisfied;
(B) w is a modular on X iff conditions (i$_\Leftarrow$) and (iii) are satisfied;
(C) w is a strict modular on X iff (i$_s$) and (iii) are satisfied.

Replacing (iii) by (iv) in the right-hand sides of (A)–(C), we get the characterization of w to be a convex pseudomodular / modular / strict modular.

Properties (i$_\Leftarrow$), (i$_s$), (iii), and (iv) of w from (1.3.1) are expressed as follows.

Proposition 1.3.1. (a) (i$_\Leftarrow$) *is equivalent to* $g \not\equiv 0$;
(b) (i$_s$) *if and only if* $g(\lambda) \neq 0$ *for all* $\lambda > 0$;
(c) (iii) *if and only if* g *is nonincreasing on* $(0, \infty)$;
(d) (iv) *if and only if* $\lambda \mapsto \lambda g(\lambda)$ *is nonincreasing on* $(0, \infty)$.

Proof. Let $\lambda, \mu > 0$ and $x, y, z \in X$.

(a) ($\Rightarrow$) To show that $g \not\equiv 0$, choose $x \neq y$, so that $d(x, y) \neq 0$. If $g \equiv 0$, then $w_\lambda(x, y) = 0$ for all $\lambda > 0$, and, by virtue of (i$_\Leftarrow$), $x = y$, which is a contradiction.
   ($\Leftarrow$) Let $w_\lambda(x, y) = 0$ for all $\lambda > 0$. Since $g \not\equiv 0$, there exists $\lambda_0 > 0$ such that $g(\lambda_0) \neq 0$, and since $g(\lambda_0)d(x, y) = 0$, we have $d(x, y) = 0$, and so, $x = y$.
(b) If $x \neq y$, then $d(x, y) \neq 0$, and so, $w_\lambda(x, y) = g(\lambda)d(x, y) \neq 0$ for all $\lambda > 0$ if and only if $g(\lambda) \neq 0$ for all $\lambda > 0$.
(c) ($\Rightarrow$) By (iii) for w as above, $g(\lambda + \mu)d(x, y) \leq g(\lambda)d(x, z) + g(\mu)d(z, y)$. Choosing $x \neq y = z$, we find $g(\lambda + \mu)d(x, y) \leq g(\lambda)d(x, y)$, i.e., $g(\lambda + \mu) \leq g(\lambda)$.
   ($\Leftarrow$) The triangle inequality (d.3) and inequality $g(\lambda + \mu) \leq g(\lambda)$ imply

$$w_{\lambda+\mu}(x, y) = g(\lambda+\mu)d(x, y) \leq g(\lambda)d(x, z)+g(\mu)d(z, y) = w_\lambda(x, z)+w_\mu(z, y).$$

(d) Apply (c) to the function $\hat{w}_\lambda(x, y) = \lambda w_\lambda(x, y)$ (see Remark 1.2.2(d)).   □

Let us consider some particular cases of modulars (1.3.1). Note that $w \equiv 0$ is the only pseudomodular on X (corresponding to $g \equiv 0$), which is not a modular on X.

Example 1.3.2. (1) Setting $g(\lambda) = 1/\lambda$ in (1.3.1), we get the convex strict modular $w_\lambda(x, y) = d(x, y)/\lambda$ (the mean velocity between x and y in time λ from Sect. 1.1), called the *canonical modular* on the metric space (X, d). Another natural strict modular $w_\lambda(x, y) = d(x, y)$ on X, corresponding to $g \equiv 1$, is nonconvex. Due to this, the canonical modular admits more adequate properties in order to 'embed' the metric space theory into the modular space theory.

More generally, $w_\lambda(x, y) = d(x, y)/\lambda^p$ $(p \geq 0)$ is a strict modular on X, which is convex if and only if $p \geq 1$ (here λ^p may be replaced by $\exp(\lambda^p) - 1$, or λe^λ, etc.).

(2) The modular w given by $w_\lambda(x, y) = d(x, y)/\lambda$ if $0 < \lambda < 1$, and $w_\lambda(x, y) = 0$ if $\lambda \geq 1$, is convex and nonstrict. If we replace the first equality by $w_\lambda(x, y) = d(x, y)$ if $0 < \lambda < 1$, then the resulting modular w on X is nonconvex and nonstrict.

(3) Given a set X, denote by δ the *discrete metric* on X (i.e., $\delta(x, y) = 0$ if $x = y$, and $\delta(x, y) = 1$ if $x \neq y$), and let $d = \delta$ in (1.3.1).

If $g \equiv \infty$, we get the *infinite modular* on X, which is strict and convex:

$$w_\lambda(x, y) = \infty \cdot \delta(x, y) = \begin{cases} 0 & \text{if } x = y, \\ \infty & \text{if } x \neq y, \end{cases} \quad \text{for all} \quad \lambda > 0.$$

Let $\lambda_0 > 0$, and $a > 0$ or $a = \infty$. Define $g(\lambda)$ by: $g(\lambda) = a$ if $0 < \lambda < \lambda_0$, and $g(\lambda) = 0$ if $\lambda \geq \lambda_0$. The *step-like modular* w on X is of the form:

$$w_\lambda(x, y) = g(\lambda) \cdot \delta(x, y) = \begin{cases} 0 & \text{if } x = y \text{ and } \lambda > 0, \\ a & \text{if } x \neq y \text{ and } 0 < \lambda < \lambda_0, \\ 0 & \text{if } x \neq y \text{ and } \lambda \geq \lambda_0. \end{cases}$$

It is nonstrict, convex if $a = \infty$, and nonconvex if $a > 0$ (is finite).

1.3.2   Families of Extended (Pseudo)metrics

A generalization of previous considerations in Sect. 1.3.1 is as follows.

Given $\lambda > 0$, let $d_\lambda : X \times X \to [0, \infty]$ be an extended pseudometric on X. Setting $w = \{w_\lambda\}_{\lambda>0}$ with $w_\lambda(x, y) = d_\lambda(x, y)$, $x, y \in X$, we find that w satisfies (i′) and (ii). So, modular classes of w on X (including convex w) are characterized as in assertions (A)–(C) of Sect. 1.3.1, where properties (i$_\Leftarrow$), (i$_s$), (iii), and (iv) are given in terms of functions $w^{x,y}$ (in place of the function g in Proposition 1.3.1):

(a) (i$_\Leftarrow$) $\iff$ $w^{x,y} \not\equiv 0$ for all $x, y \in X$ with $x \neq y$

$\iff$ condition '$d_\lambda(x, y) = 0$ for all $\lambda > 0$' implies $x = y$;

(b) (i$_s$) $\iff$ $w^{x,y}(\lambda) \neq 0$ for all $\lambda > 0$ and $x, y \in X$ with $x \neq y$

$\iff$ d_λ is an extended metric on X for all $\lambda > 0$;

(c) (iii) $\iff$ $\lambda \mapsto d_\lambda(x, y)$ is nonincreasing on $(0, \infty)$ for all $x, y \in X$;

(d) (iv) $\iff$ $\lambda \mapsto \lambda d_\lambda(x, y)$ is nonincreasing on $(0, \infty)$ for all $x, y \in X$.

Note only that in establishing assertion (c)($\Leftarrow$) we have: since $\lambda \mapsto d_\lambda(x, y)$ is nonincreasing, the triangle inequality for $w_{\lambda+\mu} = d_{\lambda+\mu}$ implies

$$w_{\lambda+\mu}(x, y) \leq d_{\lambda+\mu}(x, z) + d_{\lambda+\mu}(z, y) \leq d_\lambda(x, z) + d_\mu(z, y) = w_\lambda(x, z) + w_\mu(z, y),$$

which proves the inequality in axiom (iii).

Now, we expose two particular cases of families of (extended) metrics.

Example 1.3.3. (1) Let (X, d) be a metric space, and $h : (0, \infty) \to (0, \infty)$ be a nondecreasing function. Setting

$$w_\lambda(x, y) = \frac{d(x, y)}{h(\lambda) + d(x, y)}, \qquad \lambda > 0, \quad x, y \in X, \qquad (1.3.2)$$

we find that $w = \{w_\lambda\}_{\lambda>0}$ is a family of metrics on X such that the function $\lambda \mapsto w_\lambda(x, y)$ is nonincreasing on $(0, \infty)$, and so, by (b) and (c) above, w is a strict modular on X. For instance, the triangle inequality for w_λ is obtained as follows: the function

$$f(t) = \frac{t}{h(\lambda) + t} = 1 - \frac{h(\lambda)}{h(\lambda) + t}, \qquad t > -h(\lambda),$$

is increasing in $t \geq 0$, which together with the triangle inequality for d gives

$$w_\lambda(x, y) = f(d(x, y)) \leq f(d(x, z) + d(z, y)) = \frac{d(x, z) + d(z, y)}{h(\lambda) + d(x, z) + d(z, y)}$$

$$\leq \frac{d(x, z)}{h(\lambda) + d(x, z)} + \frac{d(z, y)}{h(\lambda) + d(z, y)} = w_\lambda(x, z) + w_\lambda(z, y).$$

The modular w is nonconvex: this is a consequence of Proposition 1.2.3(b) and the fact that $w_\lambda(x, y)$ tends to $d(x, y)/(h(+0) + d(x, y)) \leq 1$ as $\lambda \to +0$ for all $x, y \in X$.

(2) Let $T \subset [0, \infty)$, (M, d) be a metric space, and $X = M^T$ be the set of all mappings $x : T \to M$ from T into M. If w is defined by

$$w_\lambda(x, y) = \sup_{t \in T} e^{-\lambda t} d(x(t), y(t)), \qquad \lambda > 0, \quad x, y \in X,$$

then w_λ is an extended metric on X, for which the function $\lambda \mapsto w_\lambda(x, y)$ is nonincreasing on $(0, \infty)$. Hence, $w = \{w_\lambda\}_{\lambda>0}$ is a strict modular on X. Let us show that w is nonconvex. Choose $x_0, y_0 \in M$, $x_0 \neq y_0$, and set $x(t) = x_0$ and $y(t) = y_0$ for all $t \in T$. Then $x \neq y$, and $w_\lambda(x, y) = \exp(-\lambda \inf T) d(x_0, y_0)$. It follows that, as $\lambda \to +0$, we have $w_\lambda(x, y) \to d(x_0, y_0) < \infty$. It remains to refer to Proposition 1.2.3(b).

1.3.3 Classical Modulars on Real Linear Spaces

Let X be a real linear space. A functional $\rho : X \rightarrow [0, \infty]$ is said to be a *classical modular* on X in the sense of H. Nakano, J. Musielak and W. Orlicz if it satisfies the following four conditions:

(ρ.1) $\rho(0) = 0$;

(ρ.2) if $x \in X$, and $\rho(\alpha x) = 0$ for all $\alpha > 0$, then $x = 0$;

(ρ.3) $\rho(-x) = \rho(x)$ for all $x \in X$;

(ρ.4) $\rho(\alpha x + \beta y) \leq \rho(x) + \rho(y)$ for all $\alpha, \beta \geq 0$ with $\alpha + \beta = 1$, and $x, y \in X$.

If, instead of the inequality in (ρ.4), ρ satisfies

(ρ.5) $\rho(\alpha x + \beta y) \leq \alpha \rho(x) + \beta \rho(y)$,

then it is said to be a classical *convex modular* on X.

An example of a classical convex modular on X is the usual *norm* (i.e., a functional $\| \cdot \| : X \rightarrow [0, \infty)$ with properties: $\|x\| = 0 \Leftrightarrow x = 0$, $\|\alpha x\| = |\alpha| \cdot \|x\|$, and $\|x + y\| \leq \|x\| + \|y\|$ for all $x, y \in X$ and $\alpha \in \mathbb{R}$).

In the next two Propositions, we show that modulars in the sense of Definition 1.2.1 are extensions of classical modulars on linear spaces.

Proposition 1.3.4. *Given a functional $\rho : X \rightarrow [0, \infty]$, we set*

$$w_\lambda(x, y) = \rho\left(\frac{x - y}{\lambda}\right), \qquad \lambda > 0, \quad x, y \in X. \tag{1.3.3}$$

Then, we have: ρ is a classical (convex) modular on the linear space X if and only if w is a (convex) modular on the set X.

Proof. Since the assertions (ρ.1) $\Leftrightarrow$ (i$'$), (ρ.2) $\Leftrightarrow$ (i$_\Leftarrow$), and (ρ.3) $\Leftrightarrow$ (ii) are clear, we show only that (ρ.5) $\Leftrightarrow$ (iv) (the equivalence (ρ.4) $\Leftrightarrow$ (iii) is established similarly).

(ρ.5) $\Rightarrow$ (iv). Given $\lambda, \mu > 0$ and $x, y, z \in X$, we have

$$\frac{x - y}{\lambda + \mu} = \frac{\lambda}{\lambda + \mu} \cdot \frac{x - z}{\lambda} + \frac{\mu}{\lambda + \mu} \cdot \frac{z - y}{\mu} = \alpha x' + \beta y', \tag{1.3.4}$$

where

$$\alpha = \frac{\lambda}{\lambda + \mu} > 0, \quad \beta = \frac{\mu}{\lambda + \mu} > 0, \quad \alpha + \beta = 1, \quad x' = \frac{x - z}{\lambda} \quad \text{and} \quad y' = \frac{z - y}{\mu}.$$

By virtue of (1.3.3) and (ρ.5), we obtain the inequality in axiom (iv) as follows:

$$w_{\lambda + \mu}(x, y) = \rho\left(\frac{x - y}{\lambda + \mu}\right) = \rho(\alpha x' + \beta y') \leq \alpha \rho(x') + \beta \rho(y')$$

$$= \frac{\lambda}{\lambda + \mu} w_\lambda(x, z) + \frac{\mu}{\lambda + \mu} w_\mu(z, y).$$

(iv) $\Rightarrow$ (ρ.5). Assume that $\alpha > 0$, $\beta > 0$, and $\alpha + \beta = 1$ (otherwise, (ρ.5) is obvious). Taking into account (1.3.3) and (iv), for $x, y \in X$, we get

$$\rho(\alpha x + \beta y) = \rho\left(\frac{\alpha x - (-\beta y)}{\alpha + \beta}\right) = w_{\alpha+\beta}(\alpha x, -\beta y)$$

$$\leq \frac{\alpha}{\alpha + \beta} w_\alpha(\alpha x, 0) + \frac{\beta}{\alpha + \beta} w_\beta(0, -\beta y)$$

$$= \alpha\rho\left(\frac{\alpha x}{\alpha}\right) + \beta\rho\left(\frac{\beta y}{\beta}\right) = \alpha\rho(x) + \beta\rho(y). \qquad \square$$

Proposition 1.3.5. *Suppose the function* $w : (0, \infty) \times X \times X \to [0, \infty]$ *satisfies the following two conditions:*

(I) $w_\lambda(x + z, y + z) = w_\lambda(x, y)$ *for all* $\lambda > 0$ *and* $x, y, z \in X$;
(II) $w_\lambda(\mu x, 0) = w_{\lambda/\mu}(x, 0)$ *for all* $\lambda, \mu > 0$ *and* $x \in X$.

Given $x \in X$, *we set* $\rho(x) = w_1(x, 0)$. *Then, we have:*

(a) *equality (1.3.3) holds;*
(b) w *is a (convex) modular on the set* X *if and only if* ρ *is a classical (convex) modular on the real linear space* X.

Proof. (a) By virtue of assumptions (I) and (II), we find

$$w_\lambda(x, y) = w_\lambda(x - y, y - y) = w_1\left(\frac{x - y}{\lambda}, 0\right) = \rho\left(\frac{x - y}{\lambda}\right).$$

(b) As in the proof of Proposition 1.3.4, we verify only that (iv) $\Leftrightarrow$ (ρ.5).

(iv) $\Rightarrow$ (ρ.5).   Given $\alpha, \beta > 0$ with $\alpha + \beta = 1$, and $x, y \in X$, equalities (1.3.3), (I), and (II), and condition (iv) imply

$$\rho(\alpha x + \beta y) = w_1(\alpha x, -\beta y) \leq \frac{\alpha}{\alpha + \beta} w_\alpha(\alpha x, 0) + \frac{\beta}{\alpha + \beta} w_\beta(0, -\beta y)$$

$$= \alpha w_{\alpha/\alpha}(x, 0) + \beta w_{\beta/\beta}(y, 0) = \alpha\rho(x) + \beta\rho(y).$$

(ρ.5) $\Rightarrow$ (iv).   Taking into account equality (1.3.3), this is established as the corresponding implication in the proof of Proposition 1.3.4. $\square$

Remark 1.3.6. Proposition 1.3.4 provides tools for further examples of metric modulars w, generating them from classical modulars by means of formula (1.3.3). In view of Proposition 1.3.5, modulars w on a real linear space X, not satisfying conditions (I) or (II), may be nonclassical (e.g., modulars (1.3.1) and (1.3.2)).

Example 1.3.7 (the generalized Orlicz modular). Suppose (Ω, Σ, μ) is a measure space with measure μ and $\varphi : \Omega \times [0, \infty) \to [0, \infty)$ is a function satisfying the following two conditions: (a) for every $t \in \Omega$, the function $\varphi(t, \cdot) = [u \mapsto \varphi(t, u)]$

is nondecreasing and continuous on $[0, \infty)$, $\varphi(t, u) = 0$ iff $u = 0$, and $\lim_{u \to \infty} \varphi(t, u) = \infty$; (b) for all $u \geq 0$, the function $\varphi(\cdot, u) = [t \mapsto \varphi(t, u)]$ is Σ-measurable. Let X be the set of all real- (or complex-)valued functions on Ω, which are Σ-measurable and finite μ-almost everywhere (with equality μ-almost everywhere). Then, for every $x \in X$, the function $t \mapsto \varphi(t, |x(t)|)$ is Σ-measurable on Ω, and

$$\rho(x) = \int_{\Omega} \varphi(t, |x(t)|) \, d\mu \quad \text{is a classical modular on } X,$$

known as the *generalized Orlicz modular* (note that $\rho(x) = 0$ iff $x = 0$).

1.3.4   φ-Generated Modulars

Let $\varphi : [0, \infty) \to [0, \infty]$ be a nondecreasing function such that $\varphi(0) = 0$ and $\varphi \not\equiv 0$.

Given a *normed space* $(X, \| \cdot \|)$ (i.e., X is a linear space and $\| \cdot \|$ is a norm on it), the functional $\rho(x) = \varphi(\|x\|)$, $x \in X$, is a classical modular on X; in addition, ρ is a convex modular on X if and only if φ is convex on $[0, \infty)$. Since $d(x, y) = \|x - y\|$ is a metric on X, taking into account equality (1.3.3), we proceed as follows.

Proposition 1.3.8. *Let (X, d) be a metric space. Set*

$$w_\lambda(x, y) = \varphi\left(\frac{d(x, y)}{\lambda}\right), \quad \lambda > 0, \quad x, y \in X. \tag{1.3.5}$$

Then w is a modular on X. Moreover, if φ is convex, then w is a convex modular, and if $\varphi(u) \neq 0$ for all $u > 0$, then w is strict.

Proof. We shall verify some of the properties of w directly (with no reference to ρ).

To see that (i$_{\Leftarrow}$) holds, suppose $x, y \in X$, and $w_\lambda(x, y) = 0$ for all $\lambda > 0$. If $x \neq y$, then $d(x, y) > 0$, and so, given $u > 0$, setting $\lambda_u = d(x, y)/u$, we find that $\varphi(u) = \varphi(d(x, y)/\lambda_u) = w_{\lambda_u}(x, y) = 0$. Since $\varphi(0) = 0$, we have $\varphi \equiv 0$ on $[0, \infty)$, which is in contradiction with the assumption on φ. Thus, $x = y$.

In checking axiom (iii) for w, the following observation plays a key role. Given $\alpha, \beta \geq 0$, $\alpha + \beta \leq 1$, and $u_1, u_2 \geq 0$, we have

$$(\alpha + \beta) \min\{u_1, u_2\} \leq \alpha u_1 + \beta u_2 \leq \max\{u_1, u_2\},$$

and so, since φ is nondecreasing on $[0, \infty)$,

$$\varphi(\alpha u_1 + \beta u_2) \leq \max\{\varphi(u_1), \varphi(u_2)\} \leq \varphi(u_1) + \varphi(u_2). \tag{1.3.6}$$

Now, if $\lambda, \mu > 0$ and $x, y, z \in X$, the triangle inequality for d implies

$$w_{\lambda+\mu}(x, y) = \varphi\left(\frac{d(x, y)}{\lambda + \mu}\right) \leq \varphi\left(\frac{\lambda}{\lambda + \mu} \cdot \frac{d(x, z)}{\lambda} + \frac{\mu}{\lambda + \mu} \cdot \frac{d(z, y)}{\mu}\right) \qquad (1.3.7)$$

$$\leq \varphi\left(\frac{d(x, z)}{\lambda}\right) + \varphi\left(\frac{d(z, y)}{\mu}\right) = w_{\lambda}(x, z) + w_{\mu}(z, y).$$

If, in addition, φ is convex, then we proceed from (1.3.7) as follows:

$$w_{\lambda+\mu}(x, y) \leq \frac{\lambda}{\lambda + \mu} \varphi\left(\frac{d(x, z)}{\lambda}\right) + \frac{\mu}{\lambda + \mu} \varphi\left(\frac{d(z, y)}{\mu}\right)$$

$$= \frac{\lambda}{\lambda + \mu} w_{\lambda}(x, z) + \frac{\mu}{\lambda + \mu} w_{\mu}(z, y). \qquad \square \qquad (1.3.8)$$

Example 1.3.9. Let $\varphi(u) = 0$ if $0 \leq u \leq 1$, and $\varphi(u) = a$ if $u > 1$, where $a > 0$ or $a = \infty$. Then modular (1.3.5), called the $(a, 0)$-*modular*, is of the form:

$$w_{\lambda}(x, y) = \begin{cases} a & \text{if } 0 < \lambda < d(x, y), \\ 0 & \text{if } \quad \lambda \geq d(x, y). \end{cases} \qquad (1.3.9)$$

It is nonstrict, convex if $a = \infty$, and nonconvex if $a > 0$.

If $\varphi(u) = \infty$ for all $u > 0$, we get the infinite modular from Example 1.3.2(3).

Let $(M, \| \cdot \|)$ be a normed space and $X = M^{\mathbb{N}}$ the set of all sequences $x : \mathbb{N} \to M$ equipped with the componentwise operations of addition and multiplication by scalars. As usual, given $x \in X$, we set $x_n = x(n)$ for $n \in \mathbb{N}$, and so, x is also denoted by $\{x_n\}_{n=1}^{\infty} \equiv \{x_n\}$. The functional $\rho(x) = \sum_{n=1}^{\infty} \|x_n\|^p$ $(p \geq 1)$ is a classical convex modular on the linear space X.

This gives an idea to replace the function $u \mapsto u^p$, defining ρ, by the function φ as above and consider the following more general construction.

Example 1.3.10. Let (M, d) be a metric space, $X = M^{\mathbb{N}}$, and $h : [0, \infty) \to [0, \infty)$ be a superadditive function (see Appendix A.1). Define $w : (0, \infty) \times X \times X \to [0, \infty]$ by

$$w_{\lambda}(x, y) = \sum_{n=1}^{\infty} \varphi\left(\frac{d(x_n, y_n)}{h(\lambda)}\right), \qquad \lambda > 0, \quad x, y \in X. \qquad (1.3.10)$$

Then w is a modular on X. For axioms (iii) and (iv), it is to be noted only that, by virtue of (1.3.6), we have (instead of (1.3.7))

$$\varphi\left(\frac{d(x_n, y_n)}{h(\lambda + \mu)}\right) \leq \varphi\left(\frac{h(\lambda)}{h(\lambda + \mu)} \cdot \frac{d(x_n, z_n)}{h(\lambda)} + \frac{h(\mu)}{h(\lambda + \mu)} \cdot \frac{d(z_n, y_n)}{h(\mu)}\right)$$

$$\leq \varphi\left(\frac{d(x_n, z_n)}{h(\lambda)}\right) + \varphi\left(\frac{d(z_n, y_n)}{h(\mu)}\right). \qquad (1.3.11)$$

This example can be further generalized if we allow d, φ and h to depend on n, i.e., $d(x, y) = d_n(x, y)$, $\varphi(u) = \varphi_n(u)$, and $h(\lambda) = h_n(\lambda)$.

1.3.5 Pseudomodulars on the Power Set

Given a set X, we denote by $\mathscr{P}(X) \equiv 2^X$ the family of all subsets of X, also called the *power set* of X. We employ the convention that $\sup \varnothing = 0$ and $\inf \varnothing = \infty$.

Let w be a (pseudo)modular on a set X (in the sense of (i'), (i)–(iii)). Following the idea of construction of the Hausdorff distance (see Appendix A.2), we are going to introduce a pseudomodular W on the power set $\mathscr{P}(X)$, induced by w.

Given $\lambda > 0$ and nonempty sets $A, B \in \mathscr{P}(X)$, we put

$$E_\lambda(A, B) = \sup_{x \in A} \inf_{y \in B} w_\lambda(x, y) \in [0, \infty]. \tag{1.3.12}$$

Furthermore, we set

$$E_\lambda(\varnothing, B) = 0 \quad \text{for all } \lambda > 0 \text{ and } B \in \mathscr{P}(X), \tag{1.3.13}$$

and

$$E_\lambda(A, \varnothing) = \infty \quad \text{for all } \lambda > 0 \text{ and } A \in \mathscr{P}(X), \ A \neq \varnothing. \tag{1.3.14}$$

Proposition 1.3.11. *The function $E : (0, \infty) \times \mathscr{P}(X) \times \mathscr{P}(X) \to [0, \infty]$ is well-defined and has the following two properties:*

(a) $E_\lambda(A, B) = 0$ *for all* $\lambda > 0$ *and* $A \subset B \subset X$;
(b) $E_{\lambda+\mu}(A, C) \leq E_\lambda(A, B) + E_\mu(B, C)$ *for all* $\lambda, \mu > 0$ *and* $A, B, C \in \mathscr{P}(X)$.

Proof. (a) If $A = \varnothing$, then the assertion follows from (1.3.13), and if $A \neq \varnothing$, then, given $x \in A$ (so that $x \in B$), we have, by (i'), $0 \leq \inf_{y \in B} w_\lambda(x, y) \leq w_\lambda(x, x) = 0$. Since $x \in A$ is arbitrary, (1.3.12) implies $E_\lambda(A, B) = 0$.

(b) If at least one of the sets A, B, or C is empty, then we have the possibilities shown in Table 1.1.

Now, assume that A, B and C are nonempty and apply (1.3.12). Given $x \in A$, $y \in B$, and $\lambda, \mu > 0$, by virtue of (iii) for w, we have

$$\inf_{z \in C} w_{\lambda+\mu}(x, z) \leq w_{\lambda+\mu}(x, z_1) \leq w_\lambda(x, y) + w_\mu(y, z_1) \quad \text{for all } z_1 \in C. \tag{1.3.15}$$

Taking the infimum over all $z_1 \in C$, we get, for all $y \in B$,

$$\inf_{z \in C} w_{\lambda+\mu}(x, z) \leq w_\lambda(x, y) + \inf_{z_1 \in C} w_\mu(y, z_1) \leq w_\lambda(x, y) + E_\mu(B, C).$$

Table 1.1 Inequality (b) when at least one of the sets A, B, or C is empty

Sets A, B, C	$E_{\lambda+\mu}(A, C)$	$E_\lambda(A, B)$	$E_\mu(B, C)$	Apply
$A = \varnothing, B = \varnothing, C = \varnothing$	0	0	0	(1.3.13)
$A \neq \varnothing, B = \varnothing, C = \varnothing$	∞	∞	0	(1.3.14), (1.3.13)
$A = \varnothing, B \neq \varnothing, C = \varnothing$	0	0	∞	(1.3.13), (1.3.14)
$A = \varnothing, B = \varnothing, C \neq \varnothing$	0	0	0	(1.3.13)
$A \neq \varnothing, B \neq \varnothing, C = \varnothing$	∞	$\cdots$	∞	(1.3.14), (1.3.12)
$A \neq \varnothing, B = \varnothing, C \neq \varnothing$	$\cdots$	∞	0	(1.3.12), (1.3.14), (1.3.13)
$A = \varnothing, B \neq \varnothing, C \neq \varnothing$	0	0	$\cdots$	(1.3.13), (1.3.12)

Now, taking the infimum over all $y \in B$, we find, for all $x \in A$,

$$\inf_{z \in C} w_{\lambda+\mu}(x, z) \leq \inf_{y \in B} w_\lambda(x, y) + E_\mu(B, C) \leq E_\lambda(A, B) + E_\mu(B, C),$$

and it remains to take the supremum over all $x \in A$.   □

Definition 1.3.12. The function $W : (0, \infty) \times \mathscr{P}(X) \times \mathscr{P}(X) \to [0, \infty]$, defined by

$$W_\lambda(A, B) = \max\{E_\lambda(A, B), E_\lambda(B, A)\}, \quad \lambda > 0, \quad A, B \in \mathscr{P}(X),$$

has the following properties, for all $\lambda, \mu > 0$ and $A, B, C \in \mathscr{P}(X)$:

(A) $W_\lambda(A, A) = 0$;
(B) $W_\lambda(A, B) = W_\lambda(B, A)$;
(C) $W_{\lambda+\mu}(A, C) \leq W_\lambda(A, B) + W_\mu(B, C)$.

Thus, W is (only) a pseudomodular on the power set $\mathscr{P}(X)$, called the *Hausdorff pseudomodular*, induced by w.

Note that $W_\lambda(\varnothing, \varnothing) = 0$, while $W_\lambda(A, \varnothing) = \infty$ if $\varnothing \neq A \in \mathscr{P}(X)$ and $\lambda > 0$. If w is a convex (pseudo)modular on X, then applying axiom (iv) in (1.3.15) instead of (iii), we find that W is a convex pseudomodular on $\mathscr{P}(X)$.

Further properties of W will be presented below (see Theorem 2.2.13, Example 3.3.11, and Theorem 4.1.3).

1.4 Bibliographical Notes and Comments

Section 1.1. An exposition of the theory of metric spaces can be found in many monographs and textbooks, e.g., Aleksandrov [2], Copson [33], Kaplansky [51], Kolmogorov and Fomin [54], Kumaresan [57], Kuratowski [58], Schwartz [97], Shirali and Vasudeva [98] (to mention a few). A good source of metric and distance notions is a recent book by Deza and Deza [36]. The 'strong' form of the triangle inequality is due to Lindenbaum [64]. The classical reference on pseudometric spaces is Kelley's book [52]. Extended metrics, also called generalized metrics, were studied by Jung [50] and Luxemburg [67] in connection with an extension of Banach's Fixed Point Theorem from [4].

The interpretation of a modular as a generalized velocity field was initiated by Chistyakov in [26, 28].

Section 1.2. Definition 1.2.1 of (metric) modular w on a set X appeared implicitly in Chistyakov [18, 19] in connection with the studies of (bounded variation and the like) selections of set-valued mappings, and multivalued super-position operators. Explicitly and axiomatically, (pseudo)modulars were introduced in Chistyakov [22], and their main properties were established by the author in [23–25]. The strictness condition (i_s) and modular regularizations $w_{\pm 0}$ were defined in Chistyakov [28].

Section 1.3. Examples of (pseudo)modulars relevant for specific purposes are contained in [18–29]. In Sect. 1.3 and furtheron, we add some new and more general ones. An extended metric as in Example 1.3.3(2) was first defined by Bielecki [8] in order to obtain global solutions of ordinary differential equations (see also Goebel and Kirk [41, Sect. 2]).

The term *modular* on a real linear space X, extending the notion of norm, was introduced by Nakano [80, 81], where he developed the theory of modular spaces. Nakano's axioms [81, Sect. 78] of a modular $\rho : X \to [0, \infty]$ include $(\rho.1)$–$(\rho.3)$, $(\rho.5)$, and $(\rho.6)$ $\rho(x) = \sup\{\rho(\alpha x) : 0 \le \alpha < 1\}$ for all $x \in X = X_\rho^*$ (see Sect. 1.3.3 and Remark 2.3.4(1)), i.e., ρ is a left-continuous convex semimodular on X in the sense of Musielak [75, Sect. 1].

In the special case of φ-integrable functions on $[0, 1]$ and φ-summable sequences, a theory (of not necessarily convex) modulars was initiated by Mazur and Orlicz [71], and a general theory of modular spaces was developed by Musielak and Orlicz [77]. The key axiom in the nonconvex case is axiom $(\rho.4)$.

Propositions 1.3.4 and 1.3.5 are taken from Chistyakov [22, 24]. They show that our approach to (metric) modulars on arbitrary sets X is an extension of the classical approach of Nakano, Musielak and Orlicz applied to modulars on linear spaces. In particular, classical modulars are metric modulars via (1.3.3). The same situation holds for *function modulars* on linear spaces developed by Kozlowski [55].

For $h(\lambda) = \lambda$, modular (1.3.2) can be obtained from the classical nonconvex modular $\rho(x) = |x|/(1+|x|), x \in \mathbb{R}$, by means of (1.3.3) (cf. Maligranda [68, p. 8]).

In Example 1.3.7, we follow Musielak [75, Chap. II, Sect. 7].

The material of Sects. 1.3.4 and 1.3.5 is new.

Chapter 2
Metrics on Modular Spaces

Abstract In this chapter, we address the metrizability of modular spaces.

2.1 Modular Spaces

A pseudomodular w on X (cf. Fig. 1.2 on p. 5) induces an equivalence relation $\sim$ on X as follows: given $x, y \in X$,

$$x \sim y \quad \text{iff} \quad w^{x,y} \not\equiv \infty \quad \text{iff} \quad w_\lambda(x, y) < \infty \text{ for some } \lambda > 0,$$

where $\lambda = \lambda(x, y)$, possibly, depends on x and y. A modular space is any equivalence class with respect to $\sim$. More explicitly, let us fix an element $x^\circ \in X$. The set

$$X_w^* \equiv X_w^*(x^\circ) = \{x \in X : \exists \lambda = \lambda(x) > 0 \text{ such that } w_\lambda(x, x^\circ) < \infty\}$$

is called a *modular space* (around x°), and x° is called the *center* of X_w^* (x° is a representative of the equivalence class X_w^*). Note that $w^{x,y} \not\equiv \infty$ for all $x, y \in X_w^*$.

If w_{+0} and w_{-0} are the right and left regularizations of w, then (1.2.4) imply $X_{w+0}^* = X_{w-0}^* = X_w^*$.

Two more *modular spaces* (around x°) can be defined making use of other equivalence relations on X:

$$X_w^0 \equiv X_w^0(x^\circ) = \{x \in X : w_\lambda(x, x^\circ) \to 0 \text{ as } \lambda \to \infty\}$$

and

$$X_w^{\mathrm{fin}} \equiv X_w^{\mathrm{fin}}(x^\circ) = \{x \in X : w_\lambda(x, x^\circ) < \infty \text{ for all } \lambda > 0\}.$$

As above, $X_{w+0}^0 = X_{w-0}^0 = X_w^0$ and $X_w^{\mathrm{fin}} = X_{w+0}^{\mathrm{fin}} = X_{w-0}^{\mathrm{fin}} = X_w^{\mathrm{fin}}$.

Clearly, $X_w^0 \subset X_w^*$ and $X_w^{\mathrm{fin}} \subset X_w^*$ (with proper inclusions in general). However, if w is *convex*, then $X_w^0 = X_w^*$ (see Proposition 1.2.3(c)); moreover, note that this property is independent of the center x°, i.e., $X_w^0(x^\circ) = X_w^*(x^\circ)$ for all $x^\circ \in X$.

© Springer International Publishing Switzerland 2015
V.V. Chistyakov, *Metric Modular Spaces*, SpringerBriefs in Mathematics,
DOI 10.1007/978-3-319-25283-4_2

Example 2.1.1. The inclusion relations between the three modular spaces are illustrated by the modular $w_\lambda(x, y) = g(\lambda)d(x, y)$ on a metric space (X, d) from (1.3.1):

$$X_w^* = \begin{cases} \{x^\circ\} & \text{if } g \equiv \infty, \\ X & \text{if } g \not\equiv \infty, \end{cases} \qquad X_w^0 = \begin{cases} \{x^\circ\} & \text{if } \lim_{\lambda\to\infty} g(\lambda) \neq 0, \\ X & \text{if } \lim_{\lambda\to\infty} g(\lambda) = 0, \end{cases}$$

and

$$X_w^{\text{fin}} = \begin{cases} \{x^\circ\} & \text{if } g(\lambda) = \infty \text{ for some } \lambda > 0, \\ X & \text{if } g(\lambda) < \infty \text{ for all } \lambda > 0. \end{cases}$$

In particular, for modulars $w_\lambda(x, y) = d(x, y)$ (nonconvex) and $w_\lambda(x, y) = d(x, y)/\lambda$ (convex) from Example 1.3.2(a), we have

$$X_w^0 = \{x^\circ\} \subset X_w^* = X_w^{\text{fin}} = X = X_w^0 = X_w^* = X_w^{\text{fin}}.$$

In the sequel, by the *modular space* we mean the set X_w^* (the largest among the three) if not explicitly stated otherwise.

2.2 The Basic Metric

We begin by introducing the *basic (pseudo)metric* d_w^0 on the modular space X_w^*.

Theorem 2.2.1. *Let w be a (pseudo)modular on X. Set*

$$d_w^0(x, y) = \inf\{\lambda > 0 : w_\lambda(x, y) \leq \lambda\}, \qquad x, y \in X \quad (\inf \varnothing = \infty).$$

Then d_w^0 is an extended (pseudo)metric on X. Furthermore, if $x, y \in X$, $d_w^0(x, y) < \infty$ is equivalent to $x \sim y$, and so, d_w^0 is a (pseudo)metric on $X_w^* = X_w^*(x^\circ)$ (for any $x^\circ \in X$).

Proof. 1. Clearly, $d_w^0(x, y) \in [0, \infty]$, $d_w^0(x, x) = 0$, and $d_w^0(x, y) = d_w^0(y, x)$ for all $x, y \in X$. Now, suppose w is a modular on X, and $x, y \in X$ are such that $d_w^0(x, y) = 0$. The definition of d_w^0 implies $w_\mu(x, y) \leq \mu$ for all $\mu > 0$. So, for all $\lambda > 0$ and $0 < \mu < \lambda$, we have from (1.2.1): $w_\lambda(x, y) \leq w_\mu(x, y) \leq \mu \to 0$ as $\mu \to +0$. Thus $w_\lambda(x, y) = 0$ for all $\lambda > 0$, and so, by axiom (i), $x = y$.

In order to prove the triangle inequality $d_w^0(x, y) \leq d_w^0(x, z) + d_w^0(z, y)$ for all $x, y, z \in X$, we assume that $d_w^0(x, z)$ and $d_w^0(z, y)$ are finite (otherwise, the inequality is obvious). By the definition of d_w^0, given $\lambda > d_w^0(x, z)$ and $\mu > d_w^0(z, y)$, we find $w_\lambda(x, z) \leq \lambda$ and $w_\mu(z, y) \leq \mu$, and so, axiom (iii) implies

$$w_{\lambda+\mu}(x, y) \leq w_\lambda(x, z) + w_\mu(z, y) \leq \lambda + \mu.$$

It follows that $d_w^0(x, y) \leq \lambda + \mu$, and it remains to take into account the arbitrariness of λ and μ as above.

2. If $d_w^0(x, y) < \infty$, then, for any $\lambda > d_w^0(x, y)$, we have $w_\lambda(x, y) \leq \lambda < \infty$, which means that $x \sim y$. Conversely, suppose $x \sim y$, i.e., $w_\mu(x, y) < \infty$ for some $\mu > 0$. We set $\lambda = \max\{\mu, w_\mu(x, y)\}$. Since $\lambda \geq \mu$, the monotonicity (1.2.1) of w implies $w_\lambda(x, y) \leq w_\mu(x, y) \leq \lambda$, and so, $d_w^0(x, y) \leq \lambda < \infty$.
3. Given $x, y \in X_w^*$, we have $x \sim y$, and so, $d_w^0(x, y) < \infty$. By step 1, this means that d_w^0 is a (pseudo)metric on X_w^*. □

The pair (X_w^*, d_w^0), being a (pseudo)metric space generated by the (pseudo)modular w, is called a *(pseudo)metric modular space*, and we will apply this terminology if we are interested in metric properties of X_w^* with respect to d_w^0 (or some other metric induced by w). We call X_w^* the *modular space* if the main concern is its modular properties (Sects. 4.2 and 4.3), which are outside the scope of metric properties.

Example 2.2.2. Suppose $w_\lambda(x, y) = g(\lambda)d(x, y)$ is the modular from (1.3.1), where $g : (0, \infty) \to [0, \infty]$ is a nonincreasing function, $g \not\equiv 0$, and $g \not\equiv \infty$. In the examples 1–6 below, we have $X_w^* = X$, and $x, y \in X$ and $\lambda_0 > 0$ are given.

1. If $g(\lambda) = 1/\lambda^p$ ($p \geq 0$), then $d_w^0(x, y) = (d(x, y))^{1/(p+1)}$.
2. Let $g(\lambda) = 1$ if $0 < \lambda < \lambda_0$, and $g(\lambda) = 0$ if $\lambda \geq \lambda_0$. Then w is nonstrict and nonconvex, and $d_w^0(x, y) = \min\{\lambda_0, d(x, y)\}$.
3. If $g(\lambda) = 1/\lambda$ for $0 < \lambda < \lambda_0$, and $g(\lambda) = 0$ for $\lambda \geq \lambda_0$, then w is nonstrict and convex, and $d_w^0(x, y) = \min\{\lambda_0, \sqrt{d(x, y)}\}$.
4. For $g(\lambda) = \max\{1, 1/\lambda\}$, we have: w is strict and nonconvex, and d_w^0 is given by $d_w^0(x, y) = \max\{d(x, y), \sqrt{d(x, y)}\}$.
5. If $g(\lambda) = \infty$ for $0 < \lambda < \lambda_0$, and $g(\lambda) = 0$ for $\lambda \geq \lambda_0$, then w is nonstrict and convex, and $d_w^0(x, y) = \lambda_0 \delta(x, y)$, where δ is the discrete metric on X.
6. Putting $d = \delta$, for any function g as above, we have $d_w^0(x, y) = g^0 \delta(x, y)$ with $g^0 = \inf\{\lambda > 0 : g(\lambda) \leq \lambda\}$.

Remark 2.2.3. 1. If ρ is a classical modular on a real linear space X (cf. Sect. 1.3.3), the set $X_\rho = \{x \in X : \lim_{\alpha \to +0} \rho(\alpha x) = 0\}$ is called the *modular space* (with zero as its center). The modular space X_ρ is a linear subspace of X, and the functional $|\cdot|_\rho : X_\rho \to [0, \infty)$, given by $|x|_\rho = \inf\{\varepsilon > 0 : \rho(x/\varepsilon) \leq \varepsilon\}$, is an *F-norm* on X_ρ, i.e., given $x, y \in X_\rho$, it satisfies the conditions: (F.1) $|x|_\rho = 0$ iff $x = 0$; (F.2) $|-x|_\rho = |x|_\rho$; (F.3) $|x+y|_\rho \leq |x|_\rho + |y|_\rho$; and (F.4) $|c_n x_n - cx|_\rho \to 0$ as $n \to \infty$ whenever $c_n \to c$ in $\mathbb{R}$ and $|x_n - x|_\rho \to 0$ as $n \to \infty$ (where $x_n \in X_\rho$ for $n \in \mathbb{N}$). The modular space X_w^0, which is a counterpart of X_ρ, does not play that significant role in our theory as X_ρ does in the classical theory of modulars (see also Remark 2.4.3(3)).
2. Under the assumptions of Proposition 1.3.5, where X is a real linear space and $\rho(x) = w_1(x, 0)$, we also have: $X_\rho = X_w^0(0)$ is a linear subspace of X, and the functional $|x|_\rho = d_w^0(x, 0)$, $x \in X_\rho$, is an *F-norm* on X_ρ.

In Theorem 2.2.1 (and Example 2.2.2(6)), we have encountered the quantity

$$g^0 = \inf \{\lambda > 0 : g(\lambda) \leq \lambda\}, \tag{2.2.1}$$

evaluated at the nonincreasing function $g = w^{x,y} : (0, \infty) \to [0, \infty]$, which we denoted by $d_w^0(x, y) = (w^{x,y})^0$. This quantity is worth a more detailed study.

Lemma 2.2.4. *If* $g : (0, \infty) \to [0, \infty]$ *is a nonincreasing function, then* $g^0 \in [0, \infty]$, *and*

(a) $g^0 = \inf_{\lambda > 0} \max\{\lambda, g(\lambda)\}$ *(where* $\max\{\lambda, \infty\} = \infty$ *for* $\lambda > 0$*);*
(b) $g^0 < \infty$ *if and only if* $g \not\equiv \infty$ *(so,* $g^0 = \infty \Leftrightarrow g \equiv \infty$*);*
(c) $g^0 \neq 0$ *if and only if* $g \not\equiv 0$ *(so,* $g^0 = 0 \Leftrightarrow g \equiv 0$*).*

Proof. 1. Let us prove inequality ($\leq$) in (a) and implication ($\Leftarrow$) in (b). We may assume $g \not\equiv \infty$ (otherwise, (a) reads $\inf \varnothing = \infty$ and holds trivially). For each $\lambda > 0$ such that $g(\lambda) < \infty$, we set $\lambda_1 = \max\{\lambda, g(\lambda)\}$. Then $\lambda_1 \in (0, \infty)$, $g(\lambda) \leq \lambda_1$, and since $\lambda \leq \lambda_1$ and g is nonincreasing, $g(\lambda_1) \leq g(\lambda)$. So, $g(\lambda_1) \leq \lambda_1$. It follows that $g^0 \leq \lambda_1 = \max\{\lambda, g(\lambda)\}$. This proves (b)($\Leftarrow$). Taking the infimum over all $\lambda > 0$ such that $g(\lambda) < \infty$ (or over all $\lambda > 0$), we establish the inequality $g^0 \leq \ldots$ in (a).

2. Let us prove inequality ($\geq$) in (a) and implication ($\Rightarrow$) in (b). Suppose g^0 is finite. Given $\lambda_1 > g^0$, we have $g(\lambda_1) \leq \lambda_1$, and so, $g \not\equiv \infty$. This establishes (b)($\Rightarrow$). Moreover (note that the monotonicity of g is not used),

$$\inf_{\lambda > 0} \max\{\lambda, g(\lambda)\} \leq \inf_{\lambda > 0 : g(\lambda) < \infty} \max\{\lambda, g(\lambda)\} \leq \max\{\lambda_1, g(\lambda_1)\} = \lambda_1.$$

Passing to the limit as $\lambda_1 \to g^0$, we obtain the inequality $g^0 \geq \ldots$ in (a).

3. (c)($\Rightarrow$) If $g \equiv 0$, then $g^0 = \inf(0, \infty) = 0$ (equivalently, if $g^0 \neq 0$, then $g \not\equiv 0$).

    (c)($\Leftarrow$) Let $g^0 = 0$. Then $g(\mu) \leq \mu$ for all $\mu > 0$. Given $\lambda > 0$, for any $0 < \mu < \lambda$, by virtue of the monotonicity of g, we get $0 \leq g(\lambda) \leq g(\mu) \leq \mu$. Letting $\mu \to +0$, we find $g(\lambda) = 0$ for all $\lambda > 0$, i.e., $g \equiv 0$. In other words, we have shown that $g \not\equiv 0$ implies $g^0 \neq 0$. $\square$

Remark 2.2.5. It is seen from the proof of Lemma 2.2.4(a) that

$$g^0 = \inf \{\max\{\lambda, g(\lambda)\} : \lambda > 0 \text{ such that } g(\lambda) < \infty\} \in [0, \infty) \quad \text{if} \quad g \not\equiv \infty.$$

Following the same lines as in the proof of Lemma 2.2.4, it may be shown that $g^0 = \sup \{\lambda > 0 : g(\lambda) \geq \lambda\}$ (sup $\varnothing = 0$) and $g^0 = \sup_{\lambda > 0} \min\{\lambda, g(\lambda)\}$.

As a consequence of Theorem 2.2.1 and Lemma 2.2.4, we get the following

Corollary 2.2.6. $d_w^0(x, y) = \inf_{\lambda > 0} \max\{\lambda, w_\lambda(x, y)\}, \ x, y \in X.$

Given a nonincreasing function $g : (0, \infty) \to [0, \infty]$, we denote by g_{+0} and g_{-0} the right and left regularizations of g, defined (as in (1.2.2) and (1.2.3)) by:

$g_{+0}(\lambda) = g(\lambda + 0)$ and $g_{-0}(\lambda) = g(\lambda - 0)$ for all $\lambda > 0$. Functions g_{+0} and g_{-0} map $(0, \infty)$ into $[0, \infty]$ and are nonincreasing on $(0, \infty)$. Furthermore, g_{+0} is continuous from the right and g_{-0} is continuous from the left on $(0, \infty)$, and inequalities similar to (1.2.4) hold:

$$g(\lambda) \leq g(\lambda - 0) \leq g(\mu + 0) \leq g(\mu) \quad \text{in } [0, \infty] \quad \text{for all } 0 < \mu < \lambda. \tag{2.2.2}$$

Taking the above and (2.2.1) into account, we have

Lemma 2.2.7. *If* $g : (0, \infty) \to [0, \infty]$ *is nonincreasing, then* $(g_{+0})^0 = g^0 = (g_{-0})^0$.

Proof. Inequalities $(g_{+0})^0 \leq g^0 \leq (g_{-0})^0$ are consequences of the inclusions

$$\{\lambda > 0 : g(\lambda - 0) \leq \lambda\} \subset \{\lambda > 0 : g(\lambda) \leq \lambda\} \subset \{\lambda > 0 : g(\lambda + 0) \leq \lambda\},$$

which follow from (2.2.2). Now, we may assume that $g \not\equiv \infty$. Then $g_{+0} \not\equiv \infty$ and $g_{-0} \not\equiv \infty$, which ensures that g^0, $(g_{+0})^0$, and $(g_{-0})^0$ are finite.

Let us show that $g^0 \leq (g_{+0})^0$. Given $\lambda > (g_{+0})^0$, choose μ such that $(g_{+0})^0 < \mu < \lambda$. By (2.2.2) and definition of $(g_{+0})^0$, we get

$$g(\lambda) \leq g(\mu + 0) = g_{+0}(\mu) \leq \mu < \lambda.$$

Hence $g^0 \leq \lambda$. Since $\lambda > (g_{+0})^0$ is arbitrary, we find $g^0 \leq (g_{+0})^0$.

In order to show that $(g_{-0})^0 \leq g^0$, we let $\lambda > g^0$. Then, for any $\mu > 0$ such that $g^0 < \mu < \lambda$, inequalities (2.2.2) and definition of g^0 imply

$$(g_{-0})(\lambda) = g(\lambda - 0) \leq g(\mu) \leq \mu < \lambda.$$

Therefore $(g_{-0})^0 \leq \lambda$. Letting $\lambda \to g^0$, we get $(g_{-0})^0 \leq g^0$.   □

Putting, for a (pseudo)modular w on X, $g = w^{x,y}$ in Lemma 2.2.7 and noting that $g_{\pm 0} = (w_{\pm 0})^{x,y}$ and $d^0_{w_{\pm 0}}(x, y) = (g_{\pm 0})^0$, we have

Corollary 2.2.8. $d^0_{w_{+0}}(x, y) = d^0_{w_{-0}}(x, y) = d^0_w(x, y)$ *for all* $x, y \in X$.

In particular, if w and w are (pseudo)modulars on X such that $w_{+0} = \mathsf{w}_{+0}$ or $w_{-0} = \mathsf{w}_{-0}$, then $d^0_w = d^0_{\mathsf{w}}$ on $X \times X$.

We conclude that the right and left regularizations of a (pseudo)modular w on X provide no new modular spaces as compared to X^*_w, X^0_w and X^{fin}_w (cf. Sect. 2.1) and no new (pseudo)metrics as compared to d^0_w.

Yet, in Sect. 2.5, we establish the existence of continuum many (equivalent) metrics on the modular space X^*_w.

This section is continued by studying the basic metric $d^0_w(x, y)$ at the level of the map $g \mapsto g^0$, applied later to nonincreasing functions $g = w^{x,y}$. Our next lemma clarifies the definition of g^0 and Lemma 2.2.7 and, along with (2.2.1), gives a method for evaluating g^0 in terms of solutions of certain inequalities.

Lemma 2.2.9 (inequalities for g^0). *Let $g : (0, \infty) \to [0, \infty]$ be a nonincreasing function with $0 < g^0 < \infty$ (i.e., $g \not\equiv 0$ and $g \not\equiv \infty$), and $\lambda > 0$. We have:*

(a) $g^0 < \lambda$ *if and only if* $g(\lambda - 0) < \lambda$;
(b) $g^0 > \lambda$ *if and only if* $g(\lambda + 0) > \lambda$;
(c) $g^0 = \lambda$ *if and only if* $g(\lambda + 0) \leq \lambda \leq g(\lambda - 0)$.

Proof. (a)($\Rightarrow$) Suppose $g^0 < \lambda$. Given λ_1 and λ_2 such that $g^0 < \lambda_1 < \lambda_2 < \lambda$, by the monotonicity of g, $g(\lambda_2) \leq g(\lambda_1)$, and the definition of g^0 implies $g(\lambda_1) \leq \lambda_1$. Hence $g(\lambda_2) \leq \lambda_1$. Passing to the limits as $\lambda_1 \to g^0$ and $\lambda_2 \to \lambda$, we get $g(\lambda - 0) \leq g^0$, where $g^0 < \lambda$, and so, $g(\lambda - 0) < \lambda$.

(a)($\Leftarrow$) By the assumption, $g(\lambda - 0) < \lambda$, where $g(\lambda - 0) = \lim_{\mu \to \lambda - 0} g(\mu)$ and $\lambda = \lim_{\mu \to \lambda - 0} \mu$. So, there exists μ_0 with $0 < \mu_0 < \lambda$ such that $g(\mu) < \mu$ for all μ with $\mu_0 \leq \mu < \lambda$. By the definition of g^0, we find $g^0 \leq \mu$, which implies $g^0 < \lambda$.

(b)($\Rightarrow$) Let $g^0 > \lambda$. For any λ_1 and λ_2 such that $g^0 > \lambda_2 > \lambda_1 > \lambda$, we have $g(\lambda_1) \geq g(\lambda_2) > \lambda_2$, where the last inequality follows from the definition of g^0: if, on the contrary, $g(\lambda_2) \leq \lambda_2$, then $g^0 \leq \lambda_2$, which contradicts the inequality $g^0 > \lambda_2$. Therefore $g(\lambda_1) > \lambda_2$. Letting $\lambda_2 \to g^0$ and $\lambda_1 \to \lambda$, we find $g(\lambda + 0) \geq g^0 > \lambda$.

(b)($\Leftarrow$) Since $\lim_{\mu \to \lambda + 0} g(\mu) = g(\lambda + 0) > \lambda = \lim_{\mu \to \lambda + 0} \mu$, there exists $\mu_0 > \lambda$ such that $g(\mu) > \mu$ for all μ with $\lambda < \mu \leq \mu_0$. It follows that $g^0 \geq \mu$ (otherwise, if $g^0 < \mu$, then the definition of g^0 implies $g(\mu) \leq \mu$, which is a contradiction). Since $\mu > \lambda$, we get $g^0 > \lambda$.

(c) The statement in (a) is equivalent to the following:

$$g^0 \geq \lambda \ \text{ if and only if } \ g(\lambda - 0) \geq \lambda, \tag{2.2.3}$$

and the one in (b) is equivalent to the assertion:

$$g^0 \leq \lambda \ \text{ if and only if } \ g(\lambda + 0) \leq \lambda. \tag{2.2.4}$$

From these two observations, (c) follows. $\square$

Remark 2.2.10. (a) Actually, a little bit more is shown in the proof of Lemma 2.2.9: $g^0 < \lambda \Rightarrow g(\lambda - 0) \leq g^0 < \lambda$ in (a), and $g^0 > \lambda \Rightarrow g(\lambda + 0) \geq g^0 > \lambda$ in (b).
(b) We have $g^0 = \inf\{\lambda > 0 : g(\lambda) < \lambda\} \equiv g^{0\prime}$ (cf. (2.2.1) and Lemma 2.2.4). In fact, this is clear if $g \equiv 0$ or $g \equiv \infty$, so let $0 < g^0 < \infty$. Since $\{\lambda > 0 : g(\lambda) < \lambda\} \subset \{\lambda > 0 : g(\lambda) \leq \lambda\}$, we get $g^0 \leq g^{0\prime}$. Now, given $\lambda > g^0$, inequalities (2.2.2) and Lemma 2.2.9(a) imply $g(\lambda) \leq g(\lambda - 0) < \lambda$, and so, $g^{0\prime} \leq \lambda$, which yields $g^{0\prime} \leq g^0$.
(c) Assuming one-sided continuity of g on $(0, \infty)$, in view of (2.2.4) and (2.2.3), we get some useful particular cases of Lemma 2.2.9:

$g^0 \leq \lambda \Leftrightarrow g(\lambda) \leq \lambda$, provided g is continuous from the right;
$g^0 < \lambda \Leftrightarrow g(\lambda) < \lambda$, provided g is continuous from the left;
$g^0 = \lambda \Leftrightarrow g(\lambda) = \lambda$ (i.e., λ is a fixed point of g), provided g is continuous.

(d) To illustrate Lemma 2.2.9, consider $g : (0, \infty) \to (0, \infty)$ defined by: $g(\lambda) = 3$ if $0 < \lambda < 1$, $g(\lambda) = 2$ if $\lambda = 1$, and $g(\lambda) = 0$ if $\lambda > 1$. Clearly, g is nonincreasing and $g^0 = \inf(1, \infty) = 1$. Inequalities in Lemma 2.2.9(c) are of the form:

$$g(1 + 0) = 0 < g^0 = 1 < 3 = g(1 - 0).$$

Although strict inequality $g(1 - 0) = 3 > 1 = \lambda$ holds in (2.2.3), we have $g^0 = \lambda = 1$. Similarly, $g(1 + 0) = 0 < 1 = \lambda$ in (2.2.4) and $g^0 = 1 = \lambda$.

Setting $g = w^{x,y}$ in Lemma 2.2.9 (for $x, y \in X_w^*$), we obtain the following important result for modulars w on X (cf. also Remark 2.2.10(a), (c)).

Theorem 2.2.11. *Let w be a (pseudo)modular on the set X, X_w^* be the modular space, $\lambda > 0$, and $x, y \in X_w^*$. Then we have:*

(a) *condition $d_w^0(x, y) < \lambda$ implies $w_{\lambda-0}(x, y) \leq d_w^0(x, y) < \lambda$, and conversely, condition $w_{\lambda-0}(x, y) < \lambda$ implies $d_w^0(x, y) < \lambda$;*
(b) *inequality $d_w^0(x, y) > \lambda$ implies $w_{\lambda+0}(x, y) \geq d_w^0(x, y) > \lambda$, and conversely, inequality $w_{\lambda+0}(x, y) > \lambda$ implies $d_w^0(x, y) > \lambda$;*
(c) *equality $d_w^0(x, y) = \lambda$ is equivalent to $w_{\lambda+0}(x, y) \leq \lambda \leq w_{\lambda-0}(x, y)$.*

Under the continuity assumptions on w, additional equivalences hold:

(d) *if w is continuous from the right, then $d_w^0(x, y) \leq \lambda \Leftrightarrow w_\lambda(x, y) \leq \lambda$;*
(e) *if w is continuous from the left, then $d_w^0(x, y) < \lambda \Leftrightarrow w_\lambda(x, y) < \lambda$;*
(f) *if w is continuous on $(0, \infty)$, then $d_w^0(x, y) = \lambda \Leftrightarrow w_\lambda(x, y) = \lambda$.*

The conclusions of Theorem 2.2.11 are sharp (cf. Remark 2.2.10(d) and (1.3.1)).

Example 2.2.12. Let w be given by (1.3.2) with $h(\lambda) = \lambda^p$ $(p > 0)$. Since w is continuous on $(0, \infty)$, by virtue of Theorem 2.2.11(f), the value $\lambda = d_w^0(x, y)$ with $x \neq y$ satisfies the equation $w_\lambda(x, y) = \lambda$, that is,

$$\lambda^{p+1} + d(x, y)\lambda - d(x, y) = 0. \tag{2.2.5}$$

If $p = 1$, then solving the corresponding quadratic equation, we get

$$d_w^0(x, y) = \frac{\sqrt{(d(x, y))^2 + 4d(x, y)} - d(x, y)}{2}. \tag{2.2.6}$$

For $p = 2$, the solution λ of the corresponding cubic equation (2.2.5) is given by Cardano's formula:

$$d_w^0(x, y) = \sqrt[3]{\frac{a}{2} + \sqrt{\left(\frac{a}{2}\right)^2 + \left(\frac{a}{3}\right)^3}} - \sqrt[3]{-\frac{a}{2} + \sqrt{\left(\frac{a}{2}\right)^2 + \left(\frac{a}{3}\right)^3}}, \tag{2.2.7}$$

where $a = d(x, y)$, and the square and cube roots of positive numbers have uniquely determined positive values. The solution by radicals of the fourth-order equation (for $p = 3$) can be obtained by Ferrari's method, and is left to the interested reader.

Note that, for any function h from (1.3.2), we have $d_w^0(x, y) < 1$.

In fact, if h is continuous on $(0, \infty)$, equality $w_\lambda(x, y) = \lambda$ is of the form $f(\lambda) = 0$, where $f(\lambda) = \lambda h(\lambda) - (1 - \lambda) d(x, y)$, and $\lambda h(\lambda) \to 0$ as $\lambda \to +0$. Setting $\lambda h(\lambda) = 0$ if $\lambda = 0$, we find that f is continuous on $[0, \infty)$, $f(0) = -d(x, y) < 0$ (if $x \neq y$), and $f(1) = h(1) > 0$. By the Intermediate Value Theorem, $f(\lambda) = 0$ for some $0 < \lambda < 1$, and so, $d_w^0(x, y) = \lambda < 1$.

In the general case, we first show that if there exists $\mu > 0$ such that

$$w_{\lambda - 0}(x, y) < \mu \text{ for all } \lambda > 0 \text{ and } x, y \in X, \text{ then } d_w^0(x, y) < \mu \text{ for all } x, y \in X.$$

Since $w_\lambda(x, y) \leq w_{\lambda - 0}(x, y) < \mu$, and this holds for $\lambda = \mu$, we find $d_w^0(x, y) \leq \mu$. If we assume that $d_w^0(x, y) = \mu$ (for some $x \neq y$), then, by Theorem 2.2.11(b), we have $w_\lambda(x, y) \geq w_{\lambda + 0}(x, y) > \lambda$ for all $0 < \lambda < d_w^0(x, y) = \mu$, and so, $w_{\mu - 0}(x, y)$ is equal to $\lim_{\lambda \to \mu - 0} w_\lambda(x, y) \geq \mu$, which contradicts the assumption. It remains to note that $w_{\lambda - 0}(x, y) < 1 = \mu$ for our modular w from (1.3.2).

One more example of a (pseudo)metric from Theorem 2.2.1 is given by the quantity d_W^0 on the power set $\mathscr{P}(X)$ of X, where W is the Hausdorff pseudomodular on $\mathscr{P}(X)$ induced by a (pseudo)modular w on X. There are two ways of obtaining a distance function on $\mathscr{P}(X)$ starting from w on X, namely

$$w \text{ on } X \xrightarrow{\text{Theorem 2.2.1}} d_w^0 \text{ on } X \xrightarrow{\text{Appendix A.1}} D_{d_w^0} \text{ on } \mathscr{P}(X)$$

and

$$w \text{ on } X \xrightarrow{\text{Section 1.3.5}} W \text{ on } \mathscr{P}(X) \xrightarrow{\text{Theorem 2.2.1}} d_W^0 \text{ on } \mathscr{P}(X).$$

Fortunately, the resulting distance functions $D_{d_w^0}$ and d_W^0 coincide on $\mathscr{P}(X)$ as the following theorem asserts.

Theorem 2.2.13. *Let w be a (pseudo)modular on X, $D = D_{d_w^0}$ be the Hausdorff distance on $\mathscr{P}(X)$ generated by the extended (pseudo)metric d_w^0 on X, and W be the Hausdorff pseudomodular on $\mathscr{P}(X)$ induced by w. Then*

$$d_W^0(A, B) = D(A, B) \quad \text{for all } A, B \in \mathscr{P}(X).$$

Proof. Since $d_W^0(\varnothing, \varnothing) = 0 = D(\varnothing, \varnothing)$, and $d_W^0(A, \varnothing) = \infty = D(A, \varnothing)$ for all $A \neq \varnothing$, we may assume that $A \neq \varnothing$ and $B \neq \varnothing$.

($\geq$) Suppose $d_W^0(A, B) = \inf \{\lambda > 0 : W_\lambda(A, B) \leq \lambda\}$ is finite, and $\lambda > d_W^0(A, B)$. Applying (1.2.4) and Theorem 2.2.11(a) (cf. also Remark 2.2.10(b)), we get

$$W_\lambda(A, B) = \max\{E_\lambda(A, B), E_\lambda(B, A)\} < \lambda,$$

and so, $E_\lambda(A, B) < \lambda$ and $E_\lambda(B, A) < \lambda$. By (1.3.12), we have $\inf_{y \in B} w_\lambda(x, y) < \lambda$ for all $x \in A$. So, for each $x \in A$ there exists $y_x \in B$ (depending also on λ) such that $w_\lambda(x, y_x) < \lambda$. The definition of d_w^0 gives $d_w^0(x, y_x) \leq \lambda$. Since

$$\inf_{y \in B} d_w^0(x, y) \leq d_w^0(x, y_x) \leq \lambda \quad \text{for all } x \in A,$$

we get $e(A, B) = \sup_{x \in A} \inf_{y \in B} d_w^0(x, y) \leq \lambda$. Similarly, $E_\lambda(B, A) < \lambda$ implies inequality $e(B, A) \leq \lambda$. Therefore $D(A, B) = \max\{e(A, B), e(B, A)\} \leq \lambda$ for all $\lambda > d_W^0(A, B)$, and so, $D(A, B) \leq d_W^0(A, B) < \infty$.

($\leq$) Let $D(A, B) < \infty$, and $\lambda > D(A, B)$ be arbitrary. Then $\lambda > e(A, B)$ as well as $\lambda > e(B, A)$. Inequality $\lambda > e(A, B) = \sup_{x \in A} \inf_{y \in B} d_w^0(x, y)$ implies that, given $x \in A$, $\lambda > \inf_{y \in B} d_w^0(x, y)$. So, for every $x \in A$ there exists $y_x \in B$ (also depending on λ) such that $\lambda > d_w^0(x, y_x)$. By the definition of d_w^0, we have $w_\lambda(x, y_x) \leq \lambda$. Since

$$\inf_{y \in B} w_\lambda(x, y) \leq w_\lambda(x, y_x) \leq \lambda \quad \text{for all } x \in A,$$

we find $E_\lambda(A, B) = \sup_{x \in A} \inf_{y \in B} w_\lambda(x, y) \leq \lambda$. Similarly, inequality $\lambda > e(B, A)$ implies $E_\lambda(B, A) \leq \lambda$. Hence $W_\lambda(A, B) = \max\{E_\lambda(A, B), E_\lambda(B, A)\} \leq \lambda$. The definition of d_W^0 yields $d_W^0(A, B) \leq \lambda$ for all $\lambda > D(A, B)$, and so, $d_W^0(A, B) \leq D(A, B) < \infty$. $\square$

2.3 The Basic Metric in the Convex Case

Now we treat the case when a (pseudo)modular w on X is *convex*: w gives rise to an additional (pseudo)metric on the modular space X_w^* to be studied below.

We make use of the following observation. As we have seen in Remark 1.2.2(d), the convexity of a (pseudo)modular w on X is equivalent to the fact that the function $\hat{w}_\lambda(x, y) = \lambda w_\lambda(x, y)$ is a (pseudo)modular on X. On the other hand, if a function $\hat{w}$ on $(0, \infty) \times X \times X$ is initially given, then we have: $\hat{w}$ is a (pseudo)modular on X if and only if $w_\lambda(x, y) = \hat{w}_\lambda(x, y)/\lambda$ is a convex (pseudo)modular on X.

From Sect. 2.1, we find

$$X_{\hat{w}}^0 \subset X_w^0 = X_w^* = X_{\hat{w}}^* \quad \text{and} \quad X_{\hat{w}}^{\text{fin}} = X_w^{\text{fin}} \subset X_w^* = X_{\hat{w}}^*. \tag{2.3.1}$$

By Theorem 2.2.1, $\hat{w}$ generates a (pseudo)metric on X_w^* of the form

$$d_{\hat{w}}^0(x, y) = \inf\{\lambda > 0 : \hat{w}_\lambda(x, y) \leq \lambda\} = \inf\{\lambda > 0 : w_\lambda(x, y) \leq 1\}. \tag{2.3.2}$$

The last expression is given in terms of w and is denoted by $d_w^*(x, y)$.

Properties of d_w^* are gathered in the following theorem, where Theorem 2.2.1 and Corollary 2.2.6 are applied to $\hat{w}_\lambda(x, y) = \lambda w_\lambda(x, y)$ and expressed via w.

Theorem 2.3.1. *Let w be a convex (pseudo)modular on X. Then*

$$d_w^*(x, y) \equiv \inf\{\lambda > 0 : w_\lambda(x, y) \le 1\} = \inf_{\lambda > 0} \max\{\lambda, \lambda w_\lambda(x, y)\}, \quad x, y \in X, \tag{2.3.3}$$

is an extended (pseudo)metric on X (with $d_w^*(x, y) < \infty \Leftrightarrow x \sim y$), whose restriction to the modular space X_w^* is a (pseudo)metric on X_w^*.

Furthermore, d_w^0 and d_w^* are nonlinearly equivalent in the following sense: given $x, y \in X_w^*$, we have

$$\min\{d_w^*(x, y), \sqrt{d_w^*(x, y)}\} \le d_w^0(x, y) \le \max\{d_w^*(x, y), \sqrt{d_w^*(x, y)}\}, \tag{2.3.4}$$

or, equivalently (written in a different way),

$$d_w^0(x, y) \cdot \min\{1, d_w^0(x, y)\} \le d_w^*(x, y) \le d_w^0(x, y) \cdot \max\{1, d_w^0(x, y)\}. \tag{2.3.5}$$

Only the second part of Theorem 2.3.1 is to be verified. For this, we need some precise inequalities for $d_w^* = d_{\hat{w}}^0$, which are reformulated from Theorem 2.2.11 (applied to $\hat{w}$) in terms of w and stated, for ease of reference, as

Theorem 2.3.2. *Let w be a convex (pseudo)modular on X, $\lambda > 0$, and $x, y \in X_w^*$. Then we have:*

(a) $d_w^*(x, y) < \lambda$ *implies* $w_{\lambda-0}(x, y) \le d_w^*(x, y)/\lambda < 1$, *and conversely,* $w_{\lambda-0}(x, y) < 1$ *implies* $d_w^*(x, y) < \lambda$;

(b) $d_w^*(x, y) > \lambda$ *implies* $w_{\lambda+0}(x, y) \ge d_w^*(x, y)/\lambda > 1$, *and conversely,* $w_{\lambda+0}(x, y) > 1$ *implies* $d_w^*(x, y) > \lambda$;

(c) $d_w^*(x, y) = \lambda$ *is equivalent to* $w_{\lambda+0}(x, y) \le 1 \le w_{\lambda-0}(x, y)$.

In addition, under the continuity assumptions on w, we get:

(d) $d_w^*(x, y) \le \lambda \Leftrightarrow w_\lambda(x, y) \le 1$, *provided w is continuous from the right;*

(e) $d_w^*(x, y) < \lambda \Leftrightarrow w_\lambda(x, y) < 1$, *provided w is continuous from the left;*

(f) $d_w^*(x, y) = \lambda \Leftrightarrow w_\lambda(x, y) = 1$, *provided w is continuous on $(0, \infty)$.*

Proof (of Theorem 2.3.1 (second part)). In steps 1 and 2, we show that inequalities $d_w^0(x, y) < 1$ and $d_w^*(x, y) < 1$ are equivalent, and if one of them holds, then

$$d_w^*(x, y) \le d_w^0(x, y) \le \sqrt{d_w^*(x, y)}. \tag{2.3.6}$$

Since $d_w^*(x, y) < 1$ implies $d_w^*(x, y) \le \sqrt{d_w^*(x, y)}$, inequality (2.3.6) proves (2.3.4).

1. Suppose $d_w^0(x, y) < 1$. Let us show that $d_w^*(x, y) \le d_w^0(x, y)$ (and so, $d_w^*(x, y) < 1$). In fact, for any number λ such that $d_w^0(x, y) < \lambda < 1$, the definition of d_w^0 gives

$w_\lambda(x,y) \leq \lambda < 1$, whence, by the definition of d_w^*, $d_w^*(x,y) \leq \lambda$. Passing to the limit as $\lambda \to d_w^0(x,y)$, we obtain the left-hand side inequality in (2.3.6).

2. Assume that $d_w^*(x,y) < 1$. Let us prove that $d_w^0(x,y) \leq \sqrt{d_w^*(x,y)}$, which is the right-hand side inequality in (2.3.6) (and so, $d_w^0(x,y) < 1$). Since $d_w^*(x,y) \leq \sqrt{d_w^*(x,y)} < 1$, for any λ such that $\sqrt{d_w^*(x,y)} < \lambda < 1$, inequalities (1.2.4) and, by virtue of convexity of w, Theorem 2.3.2(a) imply

$$w_\lambda(x,y) \leq w_{\lambda-0}(x,y) \leq \frac{d_w^*(x,y)}{\lambda} < \frac{\lambda^2}{\lambda} = \lambda.$$

By the definition of d_w^0, $d_w^0(x,y) \leq \lambda$. Letting λ tend to $\sqrt{d_w^*(x,y)}$, we obtain the desired inequality.

As a consequence of steps 1 and 2, inequalities $d_w^0(x,y) \geq 1$ and $d_w^*(x,y) \geq 1$ are equivalent, as well. In steps 3 and 4, we show that if one of these inequalities holds, then

$$\sqrt{d_w^*(x,y)} \leq d_w^0(x,y) \leq d_w^*(x,y). \tag{2.3.7}$$

Since $d_w^*(x,y) \geq 1$ implies $d_w^*(x,y) \geq \sqrt{d_w^*(x,y)}$, (2.3.7) establishes (2.3.4).

3. Inequality $d_w^*(x,y) \geq 1$ implies $d_w^0(x,y) \leq d_w^*(x,y)$: in fact, by the definition of d_w^*, $w_\lambda(x,y) \leq 1$ for all $\lambda > d_w^*(x,y)$, and since $\lambda > 1$, $w_\lambda(x,y) < \lambda$. From the definition of d_w^0, we get $d_w^0(x,y) \leq \lambda$. The assertion follows thanks to the arbitrariness of $\lambda > d_w^*(x,y)$.

4. Suppose $d_w^0(x,y) \geq 1$, and let us show that $\sqrt{d_w^*(x,y)} \leq d_w^0(x,y)$, which is the left-hand side inequality in (2.3.7). Given $\lambda > d_w^0(x,y)$, we have $w_\lambda(x,y) \leq \lambda$, and since $\lambda > 1$, $\lambda^2 > \lambda$. The convexity of w and (1.2.5) imply

$$w_{\lambda^2}(x,y) \leq \frac{\lambda}{\lambda^2} w_\lambda(x,y) \leq \frac{\lambda}{\lambda^2} \cdot \lambda = 1,$$

whence $d_w^*(x,y) \leq \lambda^2$. Letting λ go to $d_w^0(x,y)$, we get $d_w^*(x,y) \leq (d_w^0(x,y))^2$. □

Remark 2.3.3. 1. If w is nonconvex, the quantity $d_w^*(x,y) \in [0,\infty]$ from (2.3.3) has only two properties: $d_w^*(x,x) = 0$, and $d_w^*(x,y) = d_w^*(y,x)$. It follows from (2) in this Remark that $d_w^*(x,y) = 0 \not\Rightarrow x = y$, and from (4)—that the triangle inequality may not hold for d_w^*.

2. The convexity of w is essential for inequalities (2.3.4) and (2.3.5): modular (1.3.2) is nonconvex, and d_w^0 is a well-defined metric on X (e.g., (2.2.6) and (2.2.7)), but, since $w_\lambda(x,y) < 1$ for all $\lambda > 0$, we have $d_w^*(x,y) = 0$ for all $x, y \in X$ (and, in particular, d_w^* is not a metric on X).

3. In the proof of Theorem 2.3.1, the implications in steps 1 and 3, which are of the form $d_w^0(x,y) < 1 \Rightarrow d_w^*(x,y) \leq d_w^0(x,y)$, and $d_w^*(x,y) \geq 1 \Rightarrow d_w^0(x,y) \leq d_w^*(x,y)$, do not rely on the convexity of w and are valid for those (pseudo)modulars w, for which the quantity $d_w^*(x,y)$ is well-defined. The example in (2) above is consistent with the former implication.

4. For the modular $w_\lambda(x, y) = d(x, y)/\lambda^p$ ($p > 0$) from Example 2.2.2(1), we have $d_w^0(x, y) = (d(x, y))^{1/(p+1)}$ and $d_w^*(x, y) = (d(x, y))^{1/p}$, where we note that d_w^* is a metric on X if and only if w is convex, i.e., $p \geq 1$. So, for $p \geq 1$, setting $a = d(x, y)$, inequalities (2.3.6) and (2.3.7) assume the form:

$$a^{\frac{1}{p}} \leq a^{\frac{1}{p+1}} \leq a^{\frac{1}{2p}} \text{ if } 0 \leq a < 1, \text{ and } a^{\frac{1}{2p}} \leq a^{\frac{1}{p+1}} \leq a^{\frac{1}{p}} \text{ if } a \geq 1.$$

5. Inequalities (2.3.4) are the best possible: see Example 2.3.5(1).

Remark 2.3.4. 1. If ρ is a classical convex modular on a real linear space X (cf. Sect. 1.3.3 and Remark 2.2.3), then the modular space X_ρ coincides with the set $X_\rho^* = \{x \in X : \rho(\alpha x) < \infty$ for some $\alpha > 0\}$, and the functional $\|x\|_\rho = \inf\{\varepsilon > 0 : \rho(x/\varepsilon) \leq 1\}$ ($x \in X_\rho^*$) is a norm on $X_\rho = X_\rho^*$, which is *nonlinearly equivalent* to the F-norm $|x|_\rho$ in the same sense as in Theorem 2.3.1. Moreover, under the assumptions of Proposition 1.3.5, where X is a linear space and $\rho(x) = w_1(x, 0)$, we have: $X_\rho^* = X_w^*(0) = X_\rho$ is a linear subspace of X, and the functional $\|x\|_\rho = d_w^*(x, 0)$, $x \in X_\rho^*$, is a norm on X_ρ^*.

2. Similar to Corollary 2.2.8, if w is convex, then $d_{w+0}^* = d_{w-0}^* = d_w^*$ on $X \times X$. In fact, $(w^\wedge)_\lambda(x, y) \equiv \hat{w}_\lambda(x, y) = \lambda w_\lambda(x, y)$ is also a (pseudo)modular on X, and $(w_{\pm 0})^\wedge = (\hat{w})_{\pm 0} \equiv (w^\wedge)_{\pm 0}$, which can be seen as follows. Given $\lambda > 0$ and $x, y \in X$, (1.2.2) and (1.2.3) imply

$$\left((w_{\pm 0})^\wedge\right)_\lambda(x, y) = \lambda(w_{\pm 0})_\lambda(x, y) = \lambda w_{\lambda \pm 0}(x, y) = \lim_{\mu \to \lambda \pm 0} \mu w_\mu(x, y)$$

$$= \lim_{\mu \to \lambda \pm 0} (w^\wedge)_\mu(x, y) = (w^\wedge)_{\lambda \pm 0}(x, y) = \left((w^\wedge)_{\pm 0}\right)_\lambda(x, y).$$

By virtue of (2.3.3) and (2.3.2), $d_w^* = d_{w^\wedge}^0$, and Corollary 2.2.8 yields

$$d_{w \pm 0}^* = d_{(w_{\pm 0})^\wedge}^0 = d_{(w^\wedge)_{\pm 0}}^0 = d_{w^\wedge}^0 = d_w^*.$$

Example 2.3.5. Consider the modular $w_\lambda(x, y) = \varphi(d(x, y)/\lambda)$ from (1.3.5), where the function $\varphi : [0, \infty) \to [0, \infty]$ is nondecreasing and such that $\varphi(0) = 0$, $\varphi \not\equiv 0$, and $\varphi \not\equiv \infty$, (X, d) is a metric space, $x, y \in X_w^* = X$, and $\lambda > 0$.

1. Let $\varphi(u) = u^p$ ($p > 0$). Then w is strict, convex if $p \geq 1$, and nonconvex if $0 < p < 1$. For any $p > 0$, we have

$$d_w^0(x, y) = (d(x, y))^{p/(p+1)} \quad \text{and} \quad d_w^*(x, y) = d(x, y).$$

To show that inequalities (2.3.4) are the best possible, we note that if $p = 1$, then $d_w^0(x, y) = \sqrt{d_w^*(x, y)}$, and if $p > 1$, then (w is convex and) we find

$$d_w^0(x, y) = (d_w^*(x, y))^{p/(p+1)} \to d_w^*(x, y) \quad \text{as} \quad p \to \infty.$$

2. Let w be the $(a, 0)$-modular from (1.3.9). If $a = \infty$, then w is nonstrict and convex, and we have: $d_w^0(x, y) = d_w^*(x, y) = d(x, y)$. Now, if $a > 0$, then w is nonstrict and nonconvex, and we have: $d_w^0(x, y) = \min\{a, d(x, y)\}$, $d_w^*(x, y) = 0$ if $a \leq 1$, and $d_w^*(x, y) = d(x, y)$ if $a > 1$.
3. If $\varphi(u) = u$ for $0 \leq u \leq 1$, and $\varphi(u) = 1$ for $u > 1$, then the modular

$$w_\lambda(x, y) = 1 \text{ if } 0 < \lambda < d(x, y), \text{ and } w_\lambda(x, y) = \frac{d(x, y)}{\lambda} \text{ if } \lambda \geq d(x, y),$$

is strict and nonconvex, and $d_w^0(x, y) = \min\{1, \sqrt{d(x, y)}\}$.
4. Let $\varphi(u) = 0$ for $0 \leq u \leq 1$, and $\varphi(u) = u - 1$ for $u > 1$. We have:

$$w_\lambda(x, y) = \frac{d(x, y)}{\lambda} - 1 \text{ if } 0 < \lambda < d(x, y), \text{ and } w_\lambda(x, y) = 0 \text{ if } \lambda \geq d(x, y),$$

is nonstrict and convex, and (note that $d_w^0(x, y) < d(x, y)$ if $x \neq y$)

$$d_w^0(x, y) = \frac{\sqrt{1 + 4d(x, y)} - 1}{2} \quad \text{and} \quad d_w^*(x, y) = \frac{d(x, y)}{2}.$$

5. Suppose $\varphi(0) = 0$, $\varphi(u) = 1$ if $0 < u < 1$, and $\varphi(u) = u$ if $u \geq 1$. Given $\lambda > 0$ and $x, y \in X$, we have: $w_\lambda(x, y) = 0$ if $x = y$, and if $x \neq y$,

$$w_\lambda(x, y) = \frac{d(x, y)}{\lambda} \text{ if } 0 < \lambda \leq d(x, y), \text{ and } w_\lambda(x, y) = 1 \text{ if } \lambda > d(x, y).$$

Then the modular w is strict and nonconvex, $d_w^0(x, y) = \max\{1, \sqrt{d(x, y)}\}$ if $x \neq y$, and $d_w^0(x, y) = 0$ if $x = y$.
6. Suppose φ is given by: $\varphi(u) = u$ if $0 \leq u \leq 1$, $\varphi(u) = 1$ if $1 < u < 2$, and $\varphi(u) = u - 1$ if $u \geq 2$. The corresponding modular w is strict and nonconvex, and we have: $d_w^0(x, y) = \sqrt{d(x, y)}$ if $d(x, y) \leq 1$, $d_w^0(x, y) = 1$ if $1 < d(x, y) < 2$, and $d_w^0(x, y) = \frac{1}{2}(\sqrt{1 + 4d(x, y)} - 1)$ if $d(x, y) \geq 2$.

2.4 Modulars and Metrics on Sequence Spaces

Let (M, d) be a metric space, $X = M^\mathbb{N}$—the set of all sequences $x = \{x_n\}$ from M, and $x^\circ = \{x_n^\circ\} \subset M$—a given sequence (the center of a modular space). In this section, we study two special modulars defined on X.

1. The modular w from (1.3.10) with $\varphi(u) = u^p$ ($p > 0$) and $h(\lambda) = \lambda^q$ ($q \geq 1$) is strict and continuous, and it is convex if $p \geq 1$. The modular spaces (around x°) are given by

$$X_w^* = X_w^0 = X_w^{\text{fin}} = \left\{ x = \{x_n\} \in X : \sum_{n=1}^{\infty} (d(x_n, x_n^{\circ}))^p < \infty \right\}$$

(if $M = \mathbb{R}$ with metric $d(x, y) = |x - y|$ and $x^{\circ} = 0 = \{0\}_{n=1}^{\infty}$, then $X_w^*(0)$ is the usual space ℓ_p of all real p-summable sequences).

Let $H(\lambda) = \lambda(h(\lambda))^p = \lambda^{pq+1}$. The metric d_w^0 on X_w^* is of the form:

$$d_w^0(x, y) = H^{-1}\left(\sum_{n=1}^{\infty} (d(x_n, y_n))^p \right) = \left(\sum_{n=1}^{\infty} (d(x_n, y_n))^p \right)^{1/(pq+1)},$$

where $H^{-1}(\mu) = \mu^{1/(pq+1)}$ is the inverse function of H on $[0, \infty)$.

If $p \geq 1$, then w is convex, and we also have metric d_w^* on X_w^* of the form:

$$d_w^*(x, y) = h^{-1}\left(\left[\sum_{n=1}^{\infty} (d(x_n, y_n))^p \right]^{1/p} \right) = \left(\sum_{n=1}^{\infty} (d(x_n, y_n))^p \right)^{1/pq},$$

where $h^{-1} : [0, \infty) \to [0, \infty)$ is the inverse function of h (see Example 1.3.10, and Appendix A.1 concerning general superadditive functions h).

2. Given $\lambda > 0$ and $x = \{x_n\}$, $y = \{y_n\} \in X = M^{\mathbb{N}}$, we set

$$w_\lambda(x, y) = \sup_{n \in \mathbb{N}} \left(\frac{d(x_n, y_n)}{\lambda} \right)^{1/n}. \tag{2.4.1}$$

Proposition 2.4.1. $w = \{w_\lambda\}_{\lambda > 0}$ *is a strict nonconvex continuous modular on X.*

Proof. Axioms (i), (i$_s$), and (ii) are clear, and axiom (iii) follows from inequalities (1.3.11) with $\varphi(u) = u^{1/n}$ and $h(\lambda) = \lambda$.

In order to see that w is nonconvex, we show that $X_w^0(x^{\circ}) \neq X_w^*(x^{\circ})$ for some $x^{\circ} \in X$ (cf. Sect. 2.1). Choose any $x^{\circ} \in M$ and $x \in M$, $x \neq x^{\circ}$, and let $x^{\circ} = \{x^{\circ}\}_{n=1}^{\infty}$ and $x = \{x\}_{n=1}^{\infty}$ also denote the corresponding constant sequences from X. Given $\lambda > d(x, x^{\circ}) > 0$, we find

$$w_\lambda(x, x^{\circ}) = \sup_{n \in \mathbb{N}} \left(\frac{d(x, x^{\circ})}{\lambda} \right)^{1/n} = \lim_{n \to \infty} \left(\frac{d(x, x^{\circ})}{\lambda} \right)^{1/n} = 1,$$

and so, $x \in X_w^*(x^{\circ}) \setminus X_w^0(x^{\circ})$.

Let us show that $w_\lambda(x, y) \leq w_{\lambda+0}(x, y)$ and $w_{\lambda-0}(x, y) \leq w_\lambda(x, y)$ for all $\lambda > 0$ and $x, y \in X$, which, by virtue of inequalities (1.2.4), establish the continuity property of w. For any $n \in \mathbb{N}$ and $\mu > \lambda$, the definition of w implies

$$\left(\frac{d(x_n, y_n)}{\mu} \right)^{1/n} \leq w_\mu(x, y),$$

and so, as $\mu \to \lambda + 0$, we get

$$\left(\frac{d(x_n, y_n)}{\lambda}\right)^{1/n} \leq w_{\lambda+0}(x, y).$$

Taking the supremum over all $n \in \mathbb{N}$, we obtain the first inequality above. Now, given $\lambda, \mu > 0$, we have

$$w_\mu(x, y) = \sup_{n \in \mathbb{N}} \left(\frac{d(x_n, y_n)}{\lambda}\right)^{1/n} \cdot \left(\frac{\lambda}{\mu}\right)^{1/n} \leq w_\lambda(x, y) \cdot \sup_{n \in \mathbb{N}} \left(\lambda/\mu\right)^{1/n}$$

$$= w_\lambda(x, y) \cdot \max\{1, \lambda/\mu\}, \qquad x, y \in X. \tag{2.4.2}$$

It follows that if $0 < \mu < \lambda$, then $w_\mu(x, y) \leq w_\lambda(x, y) \cdot \lambda/\mu$, and so, passing to the limit as $\mu \to \lambda - 0$, we get $w_{\lambda-0}(x, y) \leq w_\lambda(x, y)$. $\qquad\square$

Note that (2.4.2) with $y = x^\circ$ proves that $X_w^{\mathrm{fin}}(x^\circ) = X_w^*(x^\circ)$, and establishes the following characterization of this modular space in terms of sequences $x = \{x_n\}$ and $x^\circ = \{x_n^\circ\}$ themselves:

$$x \in X_w^*(x^\circ) \quad \text{if and only if} \quad w_1(x, x^\circ) = \sup_{n \in \mathbb{N}} \left(d(x_n, x_n^\circ)\right)^{1/n} < \infty. \tag{2.4.3}$$

The modular space $X_w^0(x^\circ)$ is characterized in the following way.

Proposition 2.4.2. *Given $x \in X$, $x \in X_w^0(x^\circ)$ if and only if $\lim_{n \to \infty} \left(d(x_n, x_n^\circ)\right)^{1/n} = 0$.*

Proof. Suppose $x \in X_w^0(x^\circ)$. Then $w_\lambda(x, x^\circ) \to 0$ as $\lambda \to \infty$, and so, for each $\varepsilon > 0$ there exists $\lambda_0 = \lambda_0(\varepsilon) > 0$ such that

$$w_{\lambda_0}(x, x^\circ) = \sup_{n \in \mathbb{N}} \left(\frac{d(x_n, x_n^\circ)}{\lambda_0}\right)^{1/n} \leq \varepsilon. \tag{2.4.4}$$

This inequality is equivalent to

$$\left(d(x_n, x_n^\circ)\right)^{1/n} \leq (\lambda_0)^{1/n} \cdot \varepsilon \quad \text{for all} \quad n \in \mathbb{N}. \tag{2.4.5}$$

Passing to the limit superior as $n \to \infty$, we get

$$\limsup_{n \to \infty} \left(d(x_n, x_n^\circ)\right)^{1/n} \leq \varepsilon.$$

Due to the arbitrariness of $\varepsilon > 0$, $(d(x_n, x_n^\circ))^{1/n} \to 0$ as $n \to \infty$.

Now, assume that $(d(x_n, x_n^\circ))^{1/n} \to 0$ as $n \to \infty$. Given $\varepsilon > 0$, there exists a number $n_0 = n_0(\varepsilon) \in \mathbb{N}$ such that $(d(x_n, x_n^\circ))^{1/n} < \varepsilon$ for all $n > n_0$. Setting

$$\lambda_1(\varepsilon) = \max\{1, 1/\varepsilon^{n_0}\} \cdot \max_{1 \leq n \leq n_0} d(x_n, x_n^\circ)$$

and noting that

$$d(x_n, x_n^\circ) = \frac{d(x_n, x_n^\circ)}{\varepsilon^n} \cdot \varepsilon^n \leq \lambda_1(\varepsilon) \cdot \varepsilon^n \quad \text{for all} \quad 1 \leq n \leq n_0,$$

we obtain (2.4.5) with $\lambda_0 = \lambda_0(\varepsilon) = \max\{1, \lambda_1(\varepsilon)\}$. It follows that inequality (2.4.4) holds, whence, by virtue of (1.2.1), $w_\lambda(x, x^\circ) \leq w_{\lambda_0}(x, x^\circ) \leq \varepsilon$ for all $\lambda \geq \lambda_0$. This means that $w_\lambda(x, x^\circ) \to 0$ as $\lambda \to \infty$, i.e., $x \in X_w^0(x^\circ)$. □

The metric d_w^0 on the modular space $X_w^*(x^\circ)$ is given by

$$d_w^0(x, y) = \sup_{n \in \mathbb{N}} \left(d(x_n, y_n) \right)^{1/(n+1)}, \quad x, y \in X_w^*(x^\circ). \tag{2.4.6}$$

Recalling that w is nonconvex, we note that $d_w^*(x, y) = \sup_{n \in \mathbb{N}} d(x_n, y_n)$ is only an extended metric on $X_w^*(x^\circ)$ and X (however, d_w^* is a metric on the set of all *bounded* sequences in M; see Remark 2.4.3 below).

Writing $x = \{x_n\} \in c(x^\circ)$ if $\lim_{n \to \infty} d(x_n, x_n^\circ) = 0$, and $x = \{x_n\} \in \ell_\infty(x^\circ)$ if $\sup_{n \in \mathbb{N}} d(x_n, x_n^\circ) < \infty$, we have the following (proper) inclusion relations:

$$X_w^0(x^\circ) \subset c(x^\circ) \subset \ell_\infty(x^\circ) \subset X_w^{\text{fin}}(x^\circ) = X_w^*(x^\circ). \tag{2.4.7}$$

(Here $c(x^\circ)$ is the set of all sequences in M, which are *metrically equivalent* to $x^\circ = \{x_n^\circ\}$, and $\ell_\infty(x^\circ)$ is the set of all sequences in M, which are *bounded relative to x°*.) The first inclusion is a consequence of Proposition 2.4.2, and the third one is established as follows: if $b = \sup_{n \in \mathbb{N}} d(x_n, x_n^\circ) < \infty$, then, for all $\lambda > 0$, we have:

$$w_\lambda(x, x^\circ) = \sup_{n \in \mathbb{N}} \left(\frac{d(x_n, x_n^\circ)}{\lambda} \right)^{1/n} \leq \sup_{n \in \mathbb{N}} \left(\frac{b}{\lambda} \right)^{1/n} = \max\{1, b/\lambda\} < \infty.$$

Remark 2.4.3. 1. If $x^\circ = \{x_n^\circ\}$ is a convergent sequence in M, then every sequence $x = \{x_n\} \in c(x^\circ)$ is also convergent in M (to the limit of x°), and if x° is *bounded* in M (i.e., $\sup_{n,m \in \mathbb{N}} d(x_n^\circ, x_m^\circ) < \infty$), then every $x \in \ell_\infty(x^\circ)$ is also bounded in M.
2. In the particular case when $M = \mathbb{R}$ with metric $d(x, y) = |x - y|$ and $x^\circ = 0$ is the zero sequence, we have: $c_0 = c(0)$ is the set of all real sequences convergent to zero, and $\ell_\infty = \ell_\infty(0)$ is the set of all bounded real sequences. The following examples are illustrative (see (2.4.7)): (a) $\{1/n\} \in c_0 \setminus X_w^0(0)$; (b) $\{2^n\} \in X_w^*(0) \setminus \ell_\infty$; (c) $\{2^{-n^2}\} \in X_w^0(0)$; (d) $\{2^{n^2}\} \notin X_w^*(0)$; (e) if $x = \{n\}$, then $x \in X_w^*(0)$, $d_w^0(x, 0) = \sup_{n \in \mathbb{N}} n^{1/(n+1)} < \infty$, while $d_w^*(x, 0) = \sup_{n \in \mathbb{N}} n = \infty$.

3. The classical F-norm $|x|_\rho = d_w^0(x, 0) = \sup_{n \in \mathbb{N}} |x_n|^{1/(n+1)}$, corresponding to $\rho(x) = w_1(x, 0)$ with w from (2.4.1) and $M = \mathbb{R}$, is well-defined for $x = \{x_n\}$ from $X_\rho = X_w^0(0) \subset c_0$ and satisfies conditions (F.1)–(F.4) from Remark 2.2.3. However, on the larger modular space $X_\rho^* = X_w^*(0)$ (see Remark 2.3.4(1)), the functional $|\cdot|_\rho$ does not satisfy the *continuity* condition (F.4): for instance, if $x = \{2^{n+1}\}_{n=1}^\infty$ and $\alpha_k = 1/k$, then $x \in X_\rho^* \setminus X_\rho$ and $\alpha_k \to 0$ as $k \to \infty$, but

$$|\alpha_k x|_\rho = \sup_{n \in \mathbb{N}} (\alpha_k \cdot 2^{n+1})^{1/(n+1)} = 2 \sup_{n \in \mathbb{N}} \left(\frac{1}{k}\right)^{1/(n+1)} = 2 \quad \text{for all} \quad k \in \mathbb{N}.$$

2.5 Intermediate Metrics

In Theorem 2.2.1 and Corollary 2.2.6, we have seen two expressions for metric d_w^0 on X_w^* (see also Theorem 2.3.1 if w is convex). In this section, we define and study infinitely many metrics on the modular space X_w^*.

Theorem 2.5.1. *Let w be a (pseudo)modular on the set X. Given $0 \le \theta \le 1$ and $x, y \in X$, setting*

$$d_w^\theta(x, y) = \inf_{\lambda > 0} \left[(1 - \theta) \max\{\lambda, w_\lambda(x, y)\} + \theta(\lambda + w_\lambda(x, y)) \right], \tag{2.5.1}$$

we have: d_w^θ is an extended (pseudo)metric on X, and a (pseudo)metric on the modular space $X_w^* = X_w^*(x^\circ)$ for any $x^\circ \in X$, and the following (sharp) inequalities hold:

$$d_w^0(x, y) \le (1-\theta)d_w^0(x, y)+\theta d_w^1(x, y) \le d_w^\theta(x, y) \le d_w^1(x, y) \le 2d_w^0(x, y). \tag{2.5.2}$$

Proof. Clearly, $0 \le d_w^\theta(x, y) \le \infty$ for all $x, y \in X$ and $0 \le \theta \le 1$.

1. First, we prove our theorem for $\theta = 0$ and $\theta = 1$ simultaneously (for d_w^0, this is the second proof). Given $u, v \in [0, \infty]$, we denote by $u \oplus v$ either $\max\{u, v\}$ or $u + v$ (and $u \oplus v = \infty$ if $u = \infty$ or $v = \infty$). Then $d_w^0(x, y)$ and $d_w^1(x, y)$ are expressed by the formula:

$$d_w^\oplus(x, y) = \inf_{\lambda > 0} \lambda \oplus w_\lambda(x, y), \quad x, y \in X. \tag{2.5.3}$$

1a. If $x, y \in X_w^*$, then $d_w^\oplus(x, y) < \infty$. In fact, since $x \sim y$, there exists $\lambda_0 > 0$ such that $w_{\lambda_0}(x, y) < \infty$, and so, the set $\{\lambda \oplus w_\lambda(x, y) : \lambda > 0\} \setminus \{\infty\}$ is nonempty and bounded from below by 0 (i.e., is contained in $[0, \infty)$).

1b. Given $x \in X$, we have, by (i'), $\lambda \oplus w_\lambda(x, x) = \lambda \oplus 0 = \lambda$ for all $\lambda > 0$, and so, $d_w^\oplus(x, x) = \inf_{\lambda > 0} \lambda = 0$. Now, suppose w is a modular. Let $x, y \in X$, and $d_w^\oplus(x, y) = 0$. If we show that $w_\lambda(x, y) = 0$ for all $\lambda > 0$, then axiom (i) will imply $x = y$. On the contrary, assume that $w_{\lambda_0}(x, y) \neq 0$ for some $\lambda_0 > 0$. Given $\lambda > 0$, we have two cases: if $\lambda \geq \lambda_0$, then

$$\lambda \oplus w_\lambda(x, y) \geq \lambda \oplus 0 = \lambda \geq \lambda_0,$$

and if $\lambda < \lambda_0$, then, by the monotonicity (1.2.1) of w, we find

$$\lambda \oplus w_\lambda(x, y) \geq 0 \oplus w_\lambda(x, y) = w_\lambda(x, y) \geq w_{\lambda_0}(x, y).$$

Hence $\lambda \oplus w_\lambda(x, y) \geq \min\{\lambda_0, w_{\lambda_0}(x, y)\} \equiv \lambda_1$ for all $\lambda > 0$. By the definition of $d_w^\oplus$, we get $d_w^\oplus(x, y) \geq \lambda_1 > 0$, which contradicts the assumption.

1c. Axiom (ii) for w implies the symmetry property of $d_w^\oplus$.

1d. Let us establish the triangle inequality $d_w^\oplus(x, y) \leq d_w^\oplus(x, z) + d_w^\oplus(z, y)$ for all $x, y, z \in X$. The inequality is clear if at least one summand on the right is infinite. So, we assume that both of them are finite. By (2.5.3), given $\varepsilon > 0$, there exist $\lambda = \lambda(\varepsilon) > 0$ and $\mu = \mu(\varepsilon) > 0$ such that

$$\lambda \oplus w_\lambda(x, z) \leq d_w^\oplus(x, z) + \varepsilon \quad \text{and} \quad \mu \oplus w_\mu(z, y) \leq d_w^\oplus(z, y) + \varepsilon.$$

Since $\oplus$ is max or $+$, (2.5.3) and axiom (iii) imply

$$d_w^\oplus(x, y) \leq (\lambda + \mu) \oplus w_{\lambda+\mu}(x, y) \leq (\lambda + \mu) \oplus \big(w_\lambda(x, z) + w_\mu(z, y)\big)$$
$$(2.5.4)$$
$$\leq \big(\lambda \oplus w_\lambda(x, z)\big) + \big(\mu \oplus w_\mu(z, y)\big) \leq d_w^\oplus(x, z) + \varepsilon + d_w^\oplus(z, y) + \varepsilon.$$

It remains to take into account the arbitrariness of $\varepsilon > 0$.

2. That d_w^θ is well-defined, nondegenerate (when w is a modular), and symmetric can be proved along the same lines as in steps 1a–1c. Let us show that d_w^θ satisfies the triangle inequality. Suppose $d_w^\theta(x, z)$ and $d_w^\theta(z, y)$ are finite. Given $\varepsilon > 0$, by virtue of (2.5.1), there exist $\lambda = \lambda(\varepsilon) > 0$ and $\mu = \mu(\varepsilon) > 0$ such that

$$(1 - \theta) \max\{\lambda, w_\lambda(x, z)\} + \theta\big(\lambda + w_\lambda(x, z)\big) \leq d_w^\theta(x, z) + \varepsilon,$$
$$(1 - \theta) \max\{\mu, w_\mu(z, y)\} + \theta\big(\mu + w_\mu(z, y)\big) \leq d_w^\theta(z, y) + \varepsilon.$$

Taking into account (2.5.1), axiom (iii) and the last inequality in (2.5.4), we get:

$$d_w^\theta(x, y) \leq (1 - \theta) \max\{\lambda + \mu, w_{\lambda+\mu}(x, y)\} + \theta\big(\lambda + \mu + w_{\lambda+\mu}(x, y)\big)$$
$$\leq (1 - \theta) \max\{\lambda + \mu, w_\lambda(x, z) + w_\mu(z, y)\} + \theta\big(\lambda + \mu + w_\lambda(x, z) + w_\mu(z, y)\big)$$
$$\leq (1 - \theta)\Big[\max\{\lambda, w_\lambda(x, z)\} + \max\{\mu, w_\mu(z, y)\}\Big]$$
$$+ \theta\Big[\lambda + w_\lambda(x, z) + \mu + w_\mu(z, y)\Big]$$

$$= \left[(1 - \theta) \max\{\lambda, w_\lambda(x, z)\} + \theta(\lambda + w_\lambda(x, z)) \right]$$

$$+ \left[(1 - \theta) \max\{\mu, w_\mu(z, y)\} + \theta(\mu + w_\mu(z, y)) \right]$$

$$\leq d_w^\theta(x, z) + \varepsilon + d_w^\theta(z, y) + \varepsilon.$$

By the arbitrariness of $\varepsilon > 0$, the triangle inequality for d_w^θ follows.

3. The inequalities $\max\{u, v\} \leq u + v \leq 2\max\{u, v\}$ for $u, v \geq 0$ imply

$$d_w^0(x, y) \leq d_w^1(x, y) \leq 2d_w^0(x, y) \quad \text{for all} \quad x, y \in X. \tag{2.5.5}$$

This proves also the first and fourth inequalities in (2.5.2). Since, for any $\lambda > 0$,

$$d_w^0(x, y) \leq \max\{\lambda, w_\lambda(x, y)\} \quad \text{and} \quad d_w^1(x, y) \leq \lambda + w_\lambda(x, y),$$

we find

$$(1 - \theta)d_w^0(x, y) + \theta d_w^1(x, y) \leq (1 - \theta)\max\{\lambda, w_\lambda(x, y)\} + \theta(\lambda + w_\lambda(x, y))$$

$$\leq \lambda + w_\lambda(x, y),$$

which establishes the second and third inequalities in (2.5.2). $\qquad\qquad\square$

The sharpness of inequalities (2.5.2) is elaborated in Examples 2.5.5 and 2.5.6.

Remark 2.5.2. Not only intermediate (pseudo)metrics d_w^θ between d_w^0 and d_w^1 can be introduced as in (2.5.1): given $\alpha, \beta \geq 0$ with $\alpha + \beta \neq 0$, we set

$$d_w^{\alpha,\beta}(x, y) = \inf_{\lambda > 0} \left[\alpha \max\{\lambda, w_\lambda(x, y)\} + \beta(\lambda + w_\lambda(x, y)) \right], \quad x, y \in X.$$

In this case, we have $d_w^{\alpha,\beta}(x, y) = (\alpha + \beta)d_w^\theta(x, y)$ with $\theta = \beta/(\alpha + \beta)$.

Remark 2.5.3. Different binary operations $\oplus$ on $[0, \infty)$ can be used in formula (2.5.3) to define $d_w^\oplus(x, y)$, but then only the *generalized triangle inequality* holds:

$$d_w^\oplus(x, y) \leq C\left(d_w^\oplus(x, z) + d_w^\oplus(z, y)\right) \quad \text{with} \quad C > 1. \tag{2.5.6}$$

This can be seen as follows. Suppose $\varphi : [0, \infty) \to [0, \infty)$ is a continuous function such that $\varphi(0) = 0$, $\varphi(u) > 0$ for $u > 0$ and, for some constant $C > 1$,

$$\varphi\left(\frac{u + v}{C}\right) \leq \varphi(u) + \varphi(v) \leq \varphi(u + v) \quad \text{for all} \quad u, v \geq 0. \tag{2.5.7}$$

(Here the right-hand side inequality is the superadditivity property of φ, which is satisfied, e.g., by any convex function φ; see Appendix A.1). Denoting by φ^{-1} the inverse function of φ and setting

$$u \oplus v = \varphi^{-1}(\varphi(u) + \varphi(v)) \quad \text{for all} \quad u, v \geq 0, \tag{2.5.8}$$

we find, from (2.5.7), that

$$u \oplus v \leq u + v \leq C(u \oplus v). \tag{2.5.9}$$

For instance, if $\varphi(u) = u^p$ with $p > 1$, then $u \oplus v = (u^p + v^p)^{1/p}$, and inequalities (2.5.9) hold with sharp constant $C = 2^{1-(1/p)}$, and if $\varphi(u) = e^u - 1$, then $u \oplus v = \log(e^u + e^v - 1)$, and (2.5.9) hold with sharp constant $C = 2$. Now, in order to obtain (2.5.6), we take into account (2.5.3) and (2.5.9), and find that the right-hand side in (2.5.4) is less than or equal to

$$(\lambda + \mu) + (w_\lambda(x, z) + w_\mu(z, y)) = (\lambda + w_\lambda(x, z)) + (\mu + w_\mu(z, y))$$
$$\leq C[(\lambda \oplus w_\lambda(x, z)) + (\mu \oplus w_\mu(z, y))]$$
$$\leq C[d_w^\oplus(x, z) + \varepsilon + d_w^\oplus(z, y) + \varepsilon], \quad \varepsilon > 0.$$

The generalized triangle inequality (2.5.6) can also be obtained if, instead of $d_w^\oplus(x, y)$ from (2.5.3), we consider the quantity

$$d_w^\oplus(x, y) = \inf_{\lambda > 0} (\max\{\lambda, w_\lambda(x, y)\}) \oplus (\lambda + w_\lambda(x, y))$$

with the operation $\oplus$ on $[0, \infty)$ of the form (2.5.8).

As in Corollary 2.2.8, the right w_{+0} and left w_{-0} regularizations of w do not produce new metrics of the form (2.5.1) in the following sense.

Proposition 2.5.4. $d_{w+0}^\theta(x, y) = d_{w-0}^\theta(x, y) = d_w^\theta(x, y)$ *for all* $0 \leq \theta \leq 1$ *and* $x, y \in X$.

Proof. For instance, let us verify this for $\theta = 1$. By virtue of (1.2.4), we have

$$\lambda + w_{\lambda+0}(x, y) \leq \lambda + w_\lambda(x, y) \leq \lambda + w_{\lambda-0}(x, y) \quad \text{for all} \quad \lambda > 0,$$

whence $d_{w+0}^1(x, y) \leq d_w^1(x, y) \leq d_{w-0}^1(x, y)$.

Let us show that $d_{w+0}^1(x, y) \geq d_w^1(x, y)$. Suppose $d_{w+0}^1(x, y) < \infty$, and $u > d_{w+0}^1(x, y)$. Let $u > u_1 > d_{w+0}^1(x, y)$. By (2.5.1) with $\theta = 1$, there exists $\lambda_1 > 0$ such that

$$\lim_{\lambda \to \lambda_1 + 0} (\lambda + w_\lambda(x, y)) = \lambda_1 + w_{\lambda_1+0}(x, y) \leq u_1 < u.$$

It follows that $\lambda_2 + w_{\lambda_2}(x,y) < u$ for some $\lambda_2 > \lambda_1$, which implies

$$d_w^1(x,y) = \inf_{\lambda>0} \left(\lambda + w_\lambda(x,y)\right) \le \lambda_2 + w_{\lambda_2}(x,y) < u,$$

and it remains to pass to the limit as $u \to d_{w+0}^1(x,y)$.

Now, we show that $d_w^1(x,y) \ge d_{w-0}^1(x,y)$. Let $d_w^1(x,y) < \infty$, and $u > d_w^1(x,y)$. Choose u_1 such that $u > u_1 > d_w^1(x,y)$. By (2.5.1) with $\theta = 1$, there exists $\mu_1 > 0$ such that $\mu_1 + w_{\mu_1}(x,y) \le u_1 < u$. It follows from (1.2.4) that

$$w_{\lambda_1-0}(x,y) \le w_{\mu_1}(x,y) < u - \mu_1 \quad \text{for all} \quad \lambda_1 > \mu_1,$$

and so,

$$d_{w-0}^1(x,y) \le \lambda_1 + w_{\lambda_1-0}(x,y) < \lambda_1 + u - \mu_1.$$

Passing to the limit as $\lambda_1 \to \mu_1 + 0$, we get $d_{w-0}^1(x,y) \le u$, and it remains to take into account the arbitrariness of u as above. $\qquad\square$

Example 2.5.5 (metric d_w^1).

1. Let $w_\lambda(x,y) = \lambda^{-p}d(x,y)$ be of the form (1.3.1) with $p > 0$. By Example 2.2.2(1), $d_w^0(x,y) = (d(x,y))^{1/(p+1)}$.

   Let us calculate $d_w^1(x,y) = \inf_{\lambda>0} f(\lambda)$, where $f(\lambda) = \lambda + \lambda^{-p}d(x,y)$ (and $x \ne y$). The derivative $f'(\lambda) = 1 - p\lambda^{-p-1}d(x,y)$ vanishes at $\lambda_0 = (pd(x,y))^{1/(p+1)}$, $f'(\lambda) < 0$ if $0 < \lambda < \lambda_0$, and $f'(\lambda) > 0$ if $\lambda > \lambda_0$, and so, f attains the global minimum on $(0,\infty)$ at the point λ_0, which is equal to

$$d_w^1(x,y) = f(\lambda_0) = \gamma(p) \cdot (d(x,y))^{1/(p+1)} \quad \text{for all} \quad x,y \in X,$$

   where

$$\gamma(p) = (p+1)p^{-p/(p+1)}, \quad p > 0.$$

   Note that $1 < \gamma(p) \le 2$, $\gamma(p) = 2$ if and only if $p = 1$, and $\gamma(1/p) = \gamma(p)$. The inequalities for $\gamma(p)$ can be established directly by taking the logarithm and investigating the resulting function for extrema, or they follow from (2.5.5). In particular, if $p = 1$, the expressions for d_w^0 and d_w^1 are of the form:

$$d_w^0(x,y) = \sqrt{d(x,y)} \quad \text{and} \quad d_w^1(x,y) = 2\sqrt{d(x,y)}, \quad x,y \in X.$$

2. Formulas for d_w^0 and d_w^1 above are valid in a somewhat more general case when a (pseudo)modular w on X is *p-homogeneous* with $p > 0$ in the sense that

$$w_\lambda(x,y) = \lambda^{-p}w_1(x,y) \quad \text{for all } \lambda > 0 \text{ and } x,y \in X.$$

In this case, we have

$$d_w^0(x, y) = (w_1(x, y))^{1/(p+1)} \quad \text{and} \quad d_w^1(x, y) = \gamma(p) \cdot (w_1(x, y))^{1/(p+1)}.$$

$$(2.5.10)$$

One more example of a p-homogeneous modular w on a metric space (X, d) is given by $w_\lambda(x, y) = (d(x, y)/\lambda)^p = \lambda^{-p} w_1(x, y)$ (see Example 2.3.5(1)).

3. Given a metric space (X, d) and a convex function $\varphi : [0, \infty) \to [0, \infty)$ vanishing at zero only, we set (cf. (1.3.5))

$$w_\lambda(x, y) = \lambda \varphi\left(\frac{d(x, y)}{\lambda}\right), \qquad \lambda > 0, \quad x, y \in X.$$

Then w is a strict modular on X (cf. (1.3.8)), and since φ is increasing, continuous, and admits the continuous inverse φ^{-1}, we find

$$d_w^0(x, y) = \inf\{\lambda > 0 : \varphi(d(x, y)/\lambda) \le 1\} = d(x, y)/\varphi^{-1}(1).$$

In particular, if $\varphi(u) = u^p$ with $p > 1$, we have $d_w^0(x, y) = d(x, y)$, and taking into account that

$$w_\lambda(x, y) = \lambda\left(\frac{d(x, y)}{\lambda}\right)^p = \lambda^{-(p-1)}(d(x, y))^p = \lambda^{-(p-1)} w_1(x, y),$$

we conclude from (2.5.10) (replacing p there by $p - 1$) that

$$d_w^1(x, y) = \gamma(p - 1) \cdot (w_1(x, y))^{1/p} = p(p - 1)^{(1-p)/p} \cdot d(x, y).$$

4. Setting $w_\lambda(x, y) = e^{-\lambda} d(x, y)$ and following the same reasoning as in Example 2.5.5(1), we get

$$d_w^1(x, y) = \begin{cases} d(x, y) & \text{if } d(x, y) \le 1, \\ 1 + \log d(x, y) & \text{if } d(x, y) > 1, \end{cases} \quad x, y \in X.$$

Example 2.5.6 (metric d_w^θ). In order to be able to calculate the value $d_w^\theta(x, y)$ from (2.5.1) explicitly for all $0 \le \theta \le 1$, here once again we consider the modular $w_\lambda(x, y) = \lambda^{-p} d(x, y)$ of the form (1.3.1) with $p > 0$. Since the cases $\theta = 0$ and $\theta = 1$ were considered in Example 2.5.5(1), we are left with the case when $0 < \theta < 1$ (in calculations below, we assume that $x \ne y$).

To begin with, we note that $d_w^\theta(x, y) = \inf_{\lambda > 0} f(\theta, \lambda)$, where the function $f(\theta, \lambda)$ under the infimum sign in (2.5.1) is expressed as

$$f(\theta, \lambda) = \begin{cases} f_1(\lambda) \equiv w_\lambda(x, y) + \theta\lambda & \text{if } \lambda \le w_\lambda(x, y), \\ f_2(\lambda) \equiv \lambda + \theta w_\lambda(x, y) & \text{if } \lambda > w_\lambda(x, y), \end{cases}$$

with $f_1(\lambda) = \lambda^{-p}d(x, y) + \theta\lambda$ and $f_2(\lambda) = \lambda + \theta\lambda^{-p}d(x, y)$, and the inequality $\lambda \leq w_\lambda(x, y) = \lambda^{-p}d(x, y)$ is equivalent to $\lambda \leq \lambda_0 \equiv d_w^0(x, y) = (d(x, y))^{1/(p+1)}$. Hence

$$d_w^\theta(x, y) = \min\left\{ \inf_{0<\lambda\leq\lambda_0} f_1(\lambda), \inf_{\lambda>\lambda_0} f_2(\lambda) \right\}, \qquad (2.5.11)$$

where we note that $f_1(\lambda_0) = f_2(\lambda_0) = \lambda_0(1 + \theta)$.

The derivative $f_1'(\lambda) = -\lambda^{-p-1}pd(x, y) + \theta$ is equal to zero only at the point $\lambda_1 = \lambda_0(p/\theta)^{1/(p+1)}$, $f_1'(\lambda) < 0$ if $0 < \lambda < \lambda_1$, and $f_1'(\lambda) > 0$ if $\lambda > \lambda_1$, and so, the global minimum of f_1 on $(0, \infty)$ is attained at λ_1 and is equal to

$$f_1(\lambda_1) = \lambda_0\gamma(p)\theta^{p/(p+1)}.$$

Similarly, the derivative $f_2'(\lambda) = 1 - \lambda^{-p-1}\theta pd(x, y)$ is equal to zero at the point $\lambda_2 = \lambda_0(\theta p)^{1/(p+1)}$, $f_2'(\lambda) < 0$ for $0 < \lambda < \lambda_2$, and $f_2'(\lambda) > 0$ for $\lambda > \lambda_2$, and so, f_2 attains the global minimum on $(0, \infty)$ at λ_2, where it has the value

$$f_2(\lambda_2) = \lambda_0\gamma(p)\theta^{1/(p+1)}.$$

Given $p > 0$ and $0 \leq \theta \leq 1$, we have four cases: (I) $p \geq 1$ and $\theta \leq 1/p$; (II) $p > 1$ and $1/p < \theta$; (III) $p < 1$ and $\theta \leq p$; and (IV) $p < 1$ and $p < \theta$.

Cases (I), (III).   We have $p \geq 1 \geq \theta$ in case (I), and $p \geq \theta$ in case (III), and so, $\lambda_0 \leq \lambda_1$. Since f_1 decreases on $(0, \lambda_1]$, the value $\inf_{\lambda\leq\lambda_0} f_1(\lambda)$ is equal to $f_1(\lambda_0) = \lambda_0(1 + \theta)$. Also, we have $\theta p \leq 1$ in case (I), and $\theta p < 1$ in case (III), and so, $\lambda_2 \leq \lambda_0$. Since f_2 increases on $[\lambda_2, \infty)$, the value $\inf_{\lambda>\lambda_0} f_2(\lambda)$ is equal to $f_2(\lambda_0) = \lambda_0(1 + \theta)$. By virtue of (2.5.11), $d_w^\theta(x, y) = \lambda_0(1 + \theta)$.

Case (II).   As in case (I), since $p > 1 \geq \theta$, $\inf_{\lambda\leq\lambda_0} f_1(\lambda) = \lambda_0(1 + \theta)$. Furthermore, $\theta p > 1$ implies $\lambda_0 < \lambda_2$, where λ_2 is the point of minimum of f_2 on $[\lambda_0, \infty)$, and so,

$$\inf_{\lambda>\lambda_0} f_2(\lambda) = f_2(\lambda_2) < f_2(\lambda_0) = \lambda_0(1 + \theta) = \inf_{\lambda\leq\lambda_0} f_1(\lambda).$$

It follows from (2.5.11) that $d_w^\theta(x, y) = f_2(\lambda_2) = \lambda_0\gamma(p)\theta^{1/(p+1)}$.

Case (IV).   Inequality $p < \theta$ implies $\lambda_1 < \lambda_0$, and since λ_1 is the point of minimum of f_1 on $(0, \lambda_0]$, we find

$$\inf_{\lambda\leq\lambda_0} f_1(\lambda) = f_1(\lambda_1) < f_1(\lambda_0) = \lambda_0(1 + \theta).$$

As in case (III), since $\theta p < 1$, $\inf_{\lambda>\lambda_0} f_2(\lambda) = \lambda_0(1 + \theta)$. By (2.5.11), we conclude that $d_w^\theta(x, y) = f_1(\lambda_1) = \lambda_0\gamma(p)\theta^{p/(p+1)}$.

In this way, we have shown that

$$d_w^\theta(x,y) = (d(x,y))^{1/(p+1)} \cdot \begin{cases} 1+\theta & \text{if } 0 \le \theta \le 1/p \le 1 \text{ or} \\ & \quad 0 \le \theta \le p < 1, \\ \gamma(p)\theta^{1/(p+1)} & \text{if } 0 < 1/p < \theta \le 1, \\ \gamma(p)\theta^{p/(p+1)} & \text{if } 0 < p < \theta \le 1. \end{cases} \quad (2.5.12)$$

A few comments on this formula are in order. If $\theta = 0$ or $\theta = 1$, then it gives back the values $d_w^0(x,y)$ and $d_w^1(x,y)$ from Example 2.5.5(1). If $p > 1$ and $\theta = 1/p$ in the third line of (2.5.12), then $\gamma(p)\theta^{1/(p+1)} = 1+\theta$ (as in the first line). Similarly, if $p < 1$ and $\theta = p$ in the fourth line of (2.5.12), then $\gamma(p)\theta^{p/(p+1)} = 1 + \theta$.

Note that, for any $p > 0$ and $0 \le \theta \le 1$, we have (cf. (2.5.2))

$$(1 - \theta)d_w^0(x,y) + \theta d_w^1(x,y) = (1 - \theta + \theta\gamma(p)) \cdot (d(x,y))^{1/(p+1)}.$$

For $p \ne 1$, we have $1 < \gamma(p) < 2$, so if (a) $p > 1$ and $0 < \theta < 1$, or (b) $p < 1$ and $0 < \theta \le p$, then $1 - \theta + \theta\gamma(p) < 1 + \theta$, and so,

$$(1 - \theta)d_w^0(x,y) + \theta d_w^1(x,y) < d_w^\theta(x,y), \quad x \ne y.$$

Now, if $p = 1$, then $\gamma(p) = 2$ and $1 - \theta + \theta\gamma(p) = 1 + \theta$, which imply

$$d_w^\theta(x,y) = (1 + \theta)\sqrt{d(x,y)} = (1 - \theta)d_w^0(x,y) + \theta d_w^1(x,y) \quad \text{for all } 0 \le \theta \le 1.$$

For a convex (pseudo)modular w on X, $\hat{w}_\lambda(x,y) = \lambda w_\lambda(x,y)$ is a (pseudo)modular on X, so setting $d_w^{\theta*} = d_{\hat{w}}^\theta$ and applying Theorem 2.5.1, we get

Theorem 2.5.7. *If w is a convex (pseudo)modular on X and $0 \le \theta \le 1$, then*

$$d_w^{\theta*}(x,y) = \inf_{\lambda > 0}\left[(1 - \theta)\max\{\lambda, \lambda w_\lambda(x,y)\} + \theta(\lambda + \lambda w_\lambda(x,y))\right], \quad x,y \in X,$$

is an extended (pseudo)metric on X and a (pseudo)metric on X_w^*, and

$$d_w^*(x,y) \le (1 - \theta)d_w^*(x,y) + \theta d_w^{1*}(x,y) \le d_w^{\theta*}(x,y) \le d_w^{1*}(x,y) \le 2d_w^*(x,y),$$

where (see (2.3.3)) $d_w^*(x,y) = d_w^{0*}(x,y)$.

Remark 2.5.8. Given $0 \le \theta \le 1$, $d_w^\theta(x,y) < 1$ implies $d_w^{\theta*}(x,y) \le d_w^\theta(x,y)$. In fact, for any r such that $d_w^\theta(x,y) < r < 1$ there exists $\lambda = \lambda(r) > 0$ such that

$$(1 - \theta)\max\{\lambda, w_\lambda(x,y)\} + \theta(\lambda + w_\lambda(x,y)) \le r < 1.$$

It follows that $\lambda = (1 - \theta)\lambda + \theta\lambda < 1$,

$$\max\{\lambda, \lambda w_\lambda(x, y)\} \leq \max\{\lambda, w_\lambda(x, y)\} \quad \text{and} \quad \lambda + \lambda w_\lambda(x, y) \leq \lambda + w_\lambda(x, y),$$

and so,

$$d_w^{\theta*}(x, y) \leq (1 - \theta)\max\{\lambda, \lambda w_\lambda(x, y)\} + \theta(\lambda + \lambda w_\lambda(x, y)) \leq r.$$

It remains to pass to the limit as $r \to d_w^\theta(x, y)$.

Example 2.5.9. Let $p \geq 1$ and $w_\lambda(x, y) = (d(x, y)/\lambda)^p$ be the p-homogeneous modular from Example 2.3.5(1). Then, by Example 2.5.5(1), (2),

$$d_w^1(x, y) = \gamma(p) \cdot (d(x, y))^{p/(p+1)} \quad \text{and} \quad d_w^{1*}(x, y) = \begin{cases} d(x, y) & \text{if } p = 1, \\ \gamma(p - 1)d(x, y) & \text{if } p > 1. \end{cases}$$

2.6 Bibliographical Notes and Comments

Sections 2.1 and 2.2. Modular spaces X_w^* and X_w^0 were introduced in Chistyakov [22] and studied in [24, 25, 28]. The space X_w^0 is a counterpart of the classical modular space X_ρ defined in Musielak and Orlicz [77]; see Remark 2.2.3(1), in which the main results of [77] are briefly described. As condition (ρ.4) from Sect. 1.3.3 is crucial for defining the F-norm $|x|_\rho$ on X_ρ, axiom (iii) in Definition 1.2.1 is a proper tool to define the (pseudo)metric $d_w^0(x, y)$ on the space X_w^*, which is larger than X_w^0.

The properties of $d_w^0(x, y)$ are based on the properties of quantity g^0 from (2.2.1) (recall that $d_w^0(x, y) = (w^{x,y})^0$). This allows us to obtain an alternative expression for the (pseudo)metric $d_w^0(x, y)$ in Corollary 2.2.6.

Modular space X_w^{fin} is (natural and) new. Its role will be more clear below (see Theorem 3.3.8): some 'duality' holds between the modular spaces.

Corollary 2.2.8 was first established in Chistyakov [28].

Lemma 2.2.9 and Theorem 2.2.11 are sharp refinements of Theorem 2.10 from Chistyakov [24]. Counterparts of Theorem 2.2.11(d), (e) for classical modulars are presented in Maligranda [68, Theorem 1.4].

Theorem 2.2.13 is new.

Section 2.3. In the convex case, the results of the classical modular theory are presented in Remark 2.3.4(1). They were established by Nakano [81, Sect. 81], Musielak and Orlicz [78], and Orlicz [90] (for s-convex modulars with $0 < s \leq 1$). For Orlicz modulars (i.e., integral modulars of the form $\rho(x) = \int_\Omega \varphi(|x(t)|)d\mu$), the norm $\|x\|_\rho = \inf\{\varepsilon > 0 : \rho(x/\varepsilon) \leq 1\}$ on X_ρ^* was considered by Morse and Transue [73] and Luxemburg [66]. Note that the norm $\|x\|_\rho$ is the Minkowski functional $p_A(x) = \inf\{\varepsilon > 0 : x/\varepsilon \in A\}$ of the convex set $A = \{x : \rho(x) \leq 1\}$.

Furthermore, Musielak and Orlicz [78] proved inequalities of the form (2.3.6) and (2.3.7) for classical convex modulars ρ, and Orlicz [90] established the representation $\|x\|_\rho = \inf_{t>0} \sup\{t^{-1}, \rho(tx)t^{-1}\}$ (cf. the second equality in (2.3.3)).

The (pseudo)metric $d_w^*(x, y)$ on X_w^* was introduced in Chistyakov [22]. It is seen from the expressions for $d_w^*(x, y)$ and $\|x\|_\rho$ that $d_w^*(x, y)$ is a counterpart of the norm $\|x\|_\rho$. Interestingly, the idea of definition of $d_w^*(x, y) = (\hat{w}^{x,y})^0$ has no relation with the idea of Minkowski's functional of a convex set, and relies on g^0 from (2.2.1), however, by virtue of the 'embedding' (1.3.3), for convex modulars ρ on linear spaces, we get $\|x\|_\rho = d_w^*(x, 0)$ (see Remark 2.3.4(1)).

Section 2.4. The first modular stands for illustrative purposes—its idea is to generalize, in a straightforward way, the well-known space ℓ_p of p-summable sequences. The second modular (2.4.1), mentioned in [24, Example 3.2], is more interesting and studied in detail (see also Example 4.2.7(2)). Note that modular (2.4.1) can be obtained, via (1.3.3), from the classical modular $\rho(x) = \sup_{n\in\mathbb{N}} \sqrt[n]{|x_n|}$ for $x = \{x_n\} \in \mathbb{R}^{\mathbb{N}}$, see Rolewicz [95, Example 1.2.3].

Section 2.5. The whole material of Sect. 2.5 is new. Connections with the classical modular theory are as follows. Metric $d_w^\theta(x, y)$ from (2.5.1) for $\theta = 1$ is a counterpart of the F-norm $|x|_\rho^1 = \inf_{t>0}(1 + t\rho(tx))/t, x \in X_\rho$, from Koshi and Shimogaki [53], where inequality $|x|_\rho \leq |x|_\rho^1 \leq 2|x|_\rho$ of the form (2.5.5) was also established; here $|x|_\rho = \inf\{\varepsilon > 0 : \rho(x/\varepsilon) \leq \varepsilon\}$ is the Musielak-Orlicz F-norm.

The idea to define the operation $\oplus$ in (2.5.8) is taken from Musielak [74] and Musielak and Peetre [79] (see also Musielak [75, Sect. 3]).

The classical variant of Example 2.5.5 was elaborated in Maligranda [68, p. 4].

Metric $d_w^{\theta*}(x, y)$ from Theorem 2.5.7 for $\theta = 1$ is a counterpart of the Amemiya norm $\|x\|_\rho^1 = \inf_{t>0}(1 + \rho(tx))/t, x \in X_\rho^* = X_\rho$ (see Nakano [81, Sect. 81], Hudzik and Maligranda [48], Maligranda [68, p. 6], Musielak [75, Theorem 1.10]).

For more information about the modular theory on linear spaces and Orlicz spaces we refer to Adams [1], Kozlowski [55], Krasnosel'skiĭ and Rutickiĭ [56], Lindenstrauss and Tzafriri [65], Luxemburg [66], Maligranda [68], Musielak [75], Nakano [80, 81], Orlicz [89], Rao and Ren [92, 93], Rolewicz [95].

Chapter 3
Modular Transforms

Abstract In this chapter, we study variants of modular axioms and transformations of modulars, which preserve the modularity property. It is shown that these transforms are more flexible than the metric transforms.

3.1 Variants of Modular Axioms and Metrics

The following Proposition (and its proof) shows that new metrics, defined in it, do not lead to a greater generality as compared with metrics d_w^0 and d_w^1.

Proposition 3.1.1. *If w is a (pseudo)modular on the set X and $\varphi : [0, \infty) \to [0, \infty)$ is a superadditive function, then*

$$d_w^{0,\varphi}(x, y) \equiv \inf\{\lambda > 0 : w_\lambda(x, y) \leq \varphi(\lambda)\} = \inf_{\lambda > 0} \max\{\lambda, \varphi^{-1}(w_\lambda(x, y))\}$$

and

$$d_w^{1,\varphi}(x, y) \equiv \inf_{\lambda > 0} \left(\lambda + \varphi^{-1}(w_\lambda(x, y))\right), \qquad x, y \in X,$$

are extended (pseudo)metrics on X and (pseudo)metrics on $X_w^* = X_{\varphi^{-1} \circ w}^*$, where $(\varphi^{-1} \circ w)_\lambda(x, y) = \varphi^{-1}(w_\lambda(x, y))$ is also a (pseudo)modular on X and, moreover, $d_w^{0,\varphi}(x, y) \leq d_w^{1,\varphi}(x, y) \leq 2d_w^{0,\varphi}(x, y)$ for all $x, y \in X$.

Proof. Taking into account Theorems 2.2.1 and 2.5.1 (with $\theta = 1$) and Corollary 2.2.6, it suffices to verify only that $\varphi^{-1} \circ w$ is a (pseudo)modular on X: axioms (i'), (i), and (ii) are clear, and (iii) follows from the subadditivity of the inverse function φ^{-1}. It remains to note that

$$d_w^{0,\varphi}(x, y) = \inf\{\lambda > 0 : \varphi^{-1}(w_\lambda(x, y)) \leq \lambda\} = \inf_{\lambda > 0} \max\{\lambda, \varphi^{-1}(w_\lambda(x, y))\}. \quad \square$$

© Springer International Publishing Switzerland 2015
V.V. Chistyakov, *Metric Modular Spaces*, SpringerBriefs in Mathematics,
DOI 10.1007/978-3-319-25283-4_3

3.1.1   φ-Convex Modulars

Let $\varphi : [0, \infty) \rightarrow [0, \infty)$ be an increasing function (continuous or not) such that $\varphi(0) = 0$, $\varphi(u) > 0$ for $u > 0$, and $\varphi(\infty) = \infty$.

Definition 3.1.2. A function $w : (0, \infty) \times X \times X \rightarrow [0, \infty]$ is said to be a *φ-convex modular* on X if it safisfies axioms (i), (ii) and, instead of (iv), the inequality

$$w_{\varphi(\lambda+\mu)}(x, y) \le \frac{\lambda}{\lambda + \mu} \, w_{\varphi(\lambda)}(x, z) + \frac{\mu}{\lambda + \mu} \, w_{\varphi(\mu)}(z, y)$$

for all $\lambda, \mu > 0$ and $x, y, z \in X$. If, instead of (i), w satisfies only (i′), then it is called a *φ-convex pseudomodular*.

Clearly, w is a φ-convex (pseudo)modular on X if and only if the function $\hat{w}$, given by $\hat{w}_\lambda(x, y) = \lambda w_{\varphi(\lambda)}(x, y)$, is a (pseudo)modular on X.

It follows from Theorem 2.2.1 (cf. also (2.3.2) and Theorem 2.3.1) that

$$d_w^{*,\varphi}(x, y) = \inf\{\lambda > 0 : w_{\varphi(\lambda)}(x, y) \le 1\} = \inf_{\lambda > 0} \, \max\{\lambda, \lambda w_{\varphi(\lambda)}(x, y)\}, \quad x, y \in X,$$

is an extended (pseudo)metric on X and a (pseudo)metric on X_w^*.

Furthermore, the function $w^{x,y}$, i.e., $\lambda \mapsto w_\lambda(x, y)$, is nonincreasing on $(0, \infty)$ for all $x, y \in X$: in fact, given $0 < \mu < \lambda$, we have, by (1.2.4),

$$\lambda w_{\varphi(\lambda)}(x, y) = \hat{w}_\lambda(x, y) \le \hat{w}_\mu(x, y) = \mu w_{\varphi(\mu)}(x, y),$$

and so,

$$w_\lambda(x, y) \le \frac{\varphi^{-1}(\mu)}{\varphi^{-1}(\lambda)} \, w_\mu(x, y) \le w_\mu(x, y).$$

By Lemma 2.2.4, the quantity $d_w^0(x, y) \in [0, \infty]$ from Theorem 2.2.1 is well-defined for $x, y \in X$ and is finite for $x, y \in X_w^*$, and the following inequalities hold, which generalize inequalities (2.3.5) from Theorem 2.3.1:

$$\varphi^{-1}(a) \min\{1, a\} \le d_w^{*,\varphi}(x, y) \le \varphi^{-1}(a) \max\{1, a\} \quad \text{with} \quad a = d_w^0(x, y). \quad (3.1.1)$$

The following observation shows that the notion of a φ-convex (pseudo)modular does not give a much greater generality as compared to the notion of a convex (pseudo)modular. In fact, setting $\boldsymbol{w}_\lambda(x, y) = w_{\varphi(\lambda)}(x, y)$, or $\boldsymbol{w}_\lambda(x, y) = w_{\varphi^{-1}(\lambda)}(x, y)$, we have: w is φ-convex if and only if $\boldsymbol{w}$ is convex. This assertion is a consequence of the following (in)equalities:

$$\boldsymbol{w}_{\lambda+\mu}(x, y) = w_{\varphi(\lambda+\mu)}(x, y) \le \frac{\lambda}{\lambda + \mu} \, w_{\varphi(\lambda)}(x, z) + \frac{\mu}{\lambda + \mu} \, w_{\varphi(\mu)}(z, y)$$

$$= \frac{\lambda}{\lambda + \mu} \, \boldsymbol{w}_\lambda(x, z) + \frac{\mu}{\lambda + \mu} \, \boldsymbol{w}_\mu(z, y).$$

On the other hand, φ-convex modulars are generalizations of classical s-convex (with $0 < s \leq 1$) modulars $\rho : X \to [0, \infty]$ on a real linear space X, which, along with conditions $(\rho.1)$–$(\rho.3)$ from Sect. 1.3.3, satisfy the inequality (of generalized convexity):

$$\rho(\alpha x + \beta y) \leq \alpha^s \rho(x) + \beta^s \rho(y) \quad \text{for all } x, y \in X, \ \alpha, \beta \geq 0, \ \alpha^s + \beta^s = 1.$$

In fact, if $w_\lambda(x, y) = \rho((x - y)/\lambda)$ for all $\lambda > 0$ and $x, y \in X$ (cf. (1.3.3)), then w is a φ-convex modular on X with $\varphi(u) = u^{1/s}$: noting that

$$\frac{x - y}{(\lambda + \mu)^{1/s}} = \left(\frac{\lambda}{\lambda + \mu}\right)^{1/s} \cdot \frac{x - z}{\lambda^{1/s}} + \left(\frac{\mu}{\lambda + \mu}\right)^{1/s} \cdot \frac{z - y}{\mu^{1/s}} \equiv \alpha \cdot x' + \beta \cdot y',$$

where

$$\alpha^s + \beta^s = \frac{\lambda}{\lambda + \mu} + \frac{\mu}{\lambda + \mu} = 1,$$

we find

$$w_{\varphi(\lambda + \mu)}(x, y) = \rho\left(\frac{x - y}{(\lambda + \mu)^{1/s}}\right) = \rho(\alpha \cdot x' + \beta \cdot y') \leq \alpha^s \rho(x') + \beta^s \rho(y')$$

$$= \frac{\lambda}{\lambda + \mu} \rho\left(\frac{x - z}{\lambda^{1/s}}\right) + \frac{\mu}{\lambda + \mu} \rho\left(\frac{z - y}{\mu^{1/s}}\right)$$

$$= \frac{\lambda}{\lambda + \mu} w_{\varphi(\lambda)}(x, z) + \frac{\mu}{\lambda + \mu} w_{\varphi(\mu)}(z, y).$$

In particular, inequalities (3.1.1) for $\varphi(u) = u^{1/s}$ with $0 < s \leq 1$ are of the form:

$$\min\{a^s, a^{s+1}\} \leq d_w^{*, \varphi}(x, y) \leq \max\{a^s, a^{s+1}\} \quad \text{with} \quad a = d_w^0(x, y).$$

3.1.2 Modulars Over a Convex Cone

A further generalization of the notion of modular $w_\lambda(x, y)$ is obtained if we replace the parameter $\lambda > 0$ by an element λ from a convex cone (see Appendix A.3).

Suppose $(\Lambda, d, +, \cdot)$ is an abstract convex cone, $\dot\Lambda = \Lambda \setminus \{0\}$, and $|\lambda| = d(\lambda, 0)$ is the absolute value of $\lambda \in \Lambda$.

Definition 3.1.3. A function $w : \dot\Lambda \times X \times X \to [0, \infty]$ is said to be a *modular over* Λ on X if, given $x, y, z \in X$, the following four conditions are satisfied:

(1_Λ) $x = y$ if and only if $w_\lambda(x, y) = 0$ for all $\lambda \in \dot\Lambda$;
(2_Λ) $w_\lambda(x, y) = w_\lambda(y, x)$ for all $\lambda \in \dot\Lambda$;

(3_Λ) $w_\lambda(x, y) = w_\mu(x, y)$ for all $\lambda, \mu \in \dot{\Lambda}$ such that $|\lambda| = |\mu|$;

(4_Λ) $w_{\lambda+\mu}(x, y) \leq w_\lambda(x, z) + w_\mu(z, y)$ for all $\lambda, \mu \in \dot{\Lambda}$ with $|\lambda + \mu| = |\lambda| + |\mu|$.

If, instead of the inequality in (4_Λ), we have

$$(5_\Lambda) \quad w_{\lambda+\mu}(x, y) \leq \frac{|\lambda|}{|\lambda + \mu|} w_\lambda(x, z) + \frac{|\mu|}{|\lambda + \mu|} w_\mu(z, y),$$

then w is called a *convex modular over* Λ *on* X.

The essential property of a modular w over Λ is its *monotonicity* with respect to the first variable $\lambda \in \dot{\Lambda}$ in the following sense: if $x, y \in X$ and $\lambda, \mu \in \Lambda$ are such that $0 < |\mu| < |\lambda|$, then $w_\lambda(x, y) \leq w_\mu(x, y)$. In fact, noting that $\lambda = \lambda_1 + \mu_1$, where $\lambda_1 = (|\lambda| - |\mu|)\lambda'$, $\mu_1 = |\mu|\lambda'$, $\lambda' = \lambda/|\lambda|$, and $|\lambda'| = 1$, we find that $\lambda_1, \mu_1 \in \dot{\Lambda}$, $|\lambda_1| = |\lambda| - |\mu|$ and $|\mu_1| = |\mu|$, and so, $|\lambda_1 + \mu_1| = |\lambda| = |\lambda_1| + |\mu_1|$. Thus, conditions (4_Λ), (1_Λ), and (3_Λ) imply

$$w_\lambda(x, y) = w_{\lambda_1+\mu_1}(x, y) \leq w_{\lambda_1}(x, x) + w_{\mu_1}(x, y) = w_{\mu_1}(x, y) = w_\mu(x, y).$$

Now, it follows from Lemma 2.2.4 that the quantity

$$d_w^0(x, y) = \inf\{|\lambda| : \lambda \in \dot{\Lambda} \text{ and } w_\lambda(x, y) \leq |\lambda|\} = \inf_{\lambda \in \dot{\Lambda}} \max\{|\lambda|, w_\lambda(x, y)\},$$

defined for $x, y \in X$, is an extended metric on X and a metric on the *modular space* $X_w^*(x^\circ) = \{x \in X : w_\lambda(x, x^\circ) < \infty \text{ for some } \lambda = \lambda(x) \in \dot{\Lambda}\}$, where $x^\circ \in X$.

Every (convex) modular w on X in the sense of (i)–(iv) can be considered as a (convex) modular over Λ on X: if $\tilde{w}_\lambda(x, y) = w_{|\lambda|}(x, y)$ for all $\lambda \in \dot{\Lambda}$ and $x, y \in X$, then $\tilde{w}$ satisfies (1_Λ)–(5_Λ).

3.1.3 Complex Modulars

Let $\mathbb{C}$ be the set of all complex numbers, $|\zeta|$ denote the absolute value of $\zeta \in \mathbb{C}$, and $i = \sqrt{-1}$ be the imaginary unit.

Suppose X is a complex linear space and ρ is a classical (convex) modular on X, i.e., $\rho : X \to [0, \infty]$ satisfies conditions $(\rho.1)$–$(\rho.5)$ from Sect. 1.3.3 except that axiom $(\rho.3)$ is replaced by the condition

$(\rho.3_\mathbb{C})$ $\rho(e^{ir}x) = \rho(x)$ for all $x \in X$ and $r \in \mathbb{R}$.

Setting

$$w_\lambda(x, y) = \rho\left(\frac{x - y}{\lambda}\right) \quad \text{for all} \quad \lambda \in \dot{\mathbb{C}} = \mathbb{C} \setminus \{0\} \text{ and } x, y \in X,$$

we have: w is a (convex) modular over $\Lambda = \mathbb{C}$ on X in the sense of (1_Λ)–(5_Λ). In order to see this, we verify only conditions (3_Λ) and (5_Λ). If $\lambda, \mu \in \dot{\mathbb{C}}$ and $|\lambda| = |\mu|$, then $\lambda = \mu e^{ir}$ for some $r \in \mathbb{R}$, and so, by $(\rho.3_\mathbb{C})$,

$$w_\lambda(x, y) = \rho\left(\frac{x - y}{\lambda}\right) = \rho\left(e^{-ir}\frac{x - y}{\mu}\right) = \rho\left(\frac{x - y}{\mu}\right) = w_\mu(x, y).$$

In particular, since $|-1| = 1$, we get $w_\lambda(x, y) = w_\lambda(y, x)$.

Now, assume that $x, y, z \in X$ and $\lambda, \mu \in \hat{\mathbb{C}}$ are such that $|\lambda + \mu| = |\lambda| + |\mu|$. Taking into account (1.3.4), where, for some $r, s \in \mathbb{R}$,

$$\alpha = \frac{\lambda}{\lambda + \mu} = |\alpha|e^{ir} \quad \text{and} \quad \beta = \frac{\mu}{\lambda + \mu} = |\beta|e^{is},$$

and noting that

$$|\alpha| + |\beta| = \left|\frac{\lambda}{\lambda + \mu}\right| + \left|\frac{\mu}{\lambda + \mu}\right| = (|\lambda| + |\mu|) \cdot \left|\frac{1}{\lambda + \mu}\right| = \frac{|\lambda| + |\mu|}{|\lambda + \mu|} = 1,$$

we find from (ρ.5) that (cf. (1.3.4) one more time)

$$w_{\lambda+\mu}(x, y) = \rho\left(\frac{x - y}{\lambda + \mu}\right) = \rho(\alpha x' + \beta y') \leq |\alpha|\rho(e^{ir}x') + |\beta|\rho(e^{is}y')$$

$$= |\alpha|\rho(x') + |\beta|\rho(y') = \left|\frac{\lambda}{\lambda + \mu}\right|w_\lambda(x, z) + \left|\frac{\mu}{\lambda + \mu}\right|w_\mu(z, y),$$

which implies the inequality (5_Λ).

One more modification of the notion of modular, called an *F-modular* and based on the notion of a generalized addition operation, is presented in Chistyakov [22, 23].

3.2 Miscellaneous Transforms

Here we present further examples of (pseudo)modulars and provide methods to generate new (pseudo)modulars by means of modular transforms.

The following assertions are similar to the properties of metrics.

Let w be a (pseudo)modular on X and $g : (0, \infty) \to [0, \infty]$ be a nonincreasing function. Given $\lambda > 0$ and $x, y \in X$, we set

$$w_\lambda^s(x, y) = g(\lambda)w_\lambda(x, y) \quad \text{(g-scaling of } w\text{)}$$

and

$$w_\lambda^t(x, y) = \min\{g(\lambda), w_\lambda(x, y)\} \quad \text{(g-truncation of } w\text{)}.$$

Then w^s and w^t are pseudomodulars on X, and if, in addition, w is a modular on X and $g(\lambda) \neq 0$ for all $\lambda > 0$, then w^s and w^t are also modulars on X. If w is convex, or the function $\lambda \mapsto \lambda g(\lambda)$ is nonincreasing, then w^s is convex. In particular, $w_\lambda^\vee(x, y) = w_\lambda(x, y)/\lambda$ is *always convex*. Furthermore, if w is convex and $\lambda \mapsto \lambda g(\lambda)$ is nonincreasing, then w^t is convex (here the monotonicity of $\lambda g(\lambda)$ is essential, e.g., if $w_\lambda(x, y) = 0$ if $x = y$, and $w_\lambda(x, y) = \infty$ if $x \neq y$, then w is convex, while $w_\lambda^t(x, y) = \min\{1, w_\lambda(x, y)\}$ is not).

Clearly, the sum of two (pseudo)modulars and the maximum of two (pseudo)modulars on X are also (pseudo)modulars on X.

If $h : (0, \infty) \to (0, \infty)$ is a nondecreasing function (cf. (1.3.2)), then

$$(\lambda, x, y) \mapsto \frac{w_\lambda(x, y)}{h(\lambda) + w_\lambda(x, y)} \quad \text{is a (pseudo)modular on } X \quad (\infty/\infty = 1).$$

More generally, let $\varphi, \psi : [0, \infty) \to [0, \infty)$, where φ is superadditive and ψ is subadditive. Set $w_\lambda'(x, y) = w_{\varphi(\lambda)}(x, y)$ and $w_\lambda''(x, y) = \psi(w_\lambda(x, y))$ for $\lambda > 0$ and $x, y \in X$. Then w' and w'' are (pseudo)modulars on X (cf. Proposition 3.1.1 and Sect. 3.1.1). Note that the function ψ above, given by $\psi(u) = u/(h(\lambda) + u)$, $u \geq 0$, is concave (because its second derivative is negative), and so, it is subadditive.

3.3 Right and Left Inverses

Let w be a (pseudo)modular on the set X.

Definition 3.3.1. The *right w^+* and *left w^-* inverses of w are the functions $w^+, w^- : (0, \infty) \times X \times X \to [0, \infty]$ defined, for all $\mu > 0$ and $x, y \in X$, by the rules:

$$w_\mu^+(x, y) \equiv (w^+)_\mu(x, y) = \inf \{\lambda > 0 : w_\lambda(x, y) \leq \mu\} \quad (\inf \varnothing = \infty), \quad (3.3.1)$$

$$w_\mu^-(x, y) \equiv (w^-)_\mu(x, y) = \sup \{\lambda > 0 : w_\lambda(x, y) \geq \mu\} \quad (\sup \varnothing = 0). \quad (3.3.2)$$

The properties of w^+ and w^- are gathered in the following Theorem.

Theorem 3.3.2. *Functions w^+ and w^- are (pseudo)modulars on X such that w^+ is continuous from the right and w^- is continuous from the left on $(0, \infty)$, and the following (in)equalities hold in $[0, \infty]$, for all $\lambda, \mu > 0$ and $x, y \in X$:*

$$w_\mu^+(x, y) \leq w_\mu^-(x, y), \quad and$$

$$w_\lambda^{-+}(x, y) = w_\lambda^{++}(x, y) = w_{\lambda+0}(x, y) \leq w_\lambda(x, y) \leq w_{\lambda-0}(x, y) = w_\lambda^{--}(x, y) = w_\lambda^{+-}(x, y),$$

where $w^{-+} = (w^-)^+$, $w^{++} = (w^+)^+$, $w^{--} = (w^-)^-$, and $w^{+-} = (w^+)^-$.

Proof. 1. Let us verify only axioms (i) and (iii) for w^+ and w^-.

(i) Suppose $x, y \in X$, and $w_\mu^+(x, y) = 0$ (or $w_\mu^-(x, y) = 0$) for all $\mu > 0$. Given $\lambda > 0$, we have $\lambda > w_\mu^+(x, y)$ (or $\lambda > w_\mu^-(x, y)$), and so, (3.3.1) (or (3.3.2)) implies $w_\lambda(x, y) \leq \mu$ (or $w_\lambda(x, y) < \mu$, for, otherwise, $w_\lambda(x, y) \geq \mu$ and (3.3.2) yield $w_\mu^-(x, y) \geq \lambda$) for all $\mu > 0$. So, $w_\lambda(x, y) = 0$ for all $\lambda > 0$, which, by axiom (i) for w, means that $x = y$.

(iii) Let us show that $w_{\lambda+\mu}^\pm(x, y) \leq w_\lambda^\pm(x, z) + w_\mu^\pm(z, y)$, where we may assume that the right-hand side is finite. Given $\xi > w_\lambda^+(x, z)$ and $\eta > w_\mu^+(z, y)$

(or $\xi > w_\lambda^-(x, z)$ and $\eta > w_\mu^-(z, y)$), the definition of w^+ (or w^-) gives $w_\xi(x, z) \leq \lambda$ and $w_\eta(z, y) \leq \mu$ (or $w_\xi(x, z) < \lambda$ and $w_\eta(z, y) < \mu$), and so, by axiom (iii) for w,

$$w_{\xi+\eta}(x, y) \leq w_\xi(x, z) + w_\eta(z, y) \leq \lambda + \mu$$

(or $w_{\xi+\eta}(x, y) < \lambda + \mu$ in the case of w^-). By virtue of definition (3.3.1) (or (3.3.2)), we find $w_{\lambda+\mu}^+(x, y) \leq \xi+\eta$ (or $w_{\lambda+\mu}^-(x, y) \leq \xi+\eta$, for, otherwise, the inequality $\xi + \eta < w_{\lambda+\mu}^-(x, y)$ and (3.3.2) would imply $w_{\xi+\eta}(x, y) \geq \lambda + \mu$), and it remains to take into account the arbitrariness of ξ and η as above.

2. Since w, w^+, and w^- are (pseudo)modulars on X, functions $g(\lambda) = w_\lambda(x, y)$, $g^+(\mu) = w_\mu^+(x, y)$, and $g^-(\mu) = w_\mu^-(x, y)$ are nonincreasing in λ and μ from $(0, \infty)$ for all $x, y \in X$ (see (1.2.1)), and so, the remaining properties of w^+ and w^- from Theorem 3.3.2 follow from Lemma 3.3.4 below concerning further properties of nonincreasing functions on $(0, \infty)$.   □

Remark 3.3.3. 1. By Theorem 3.3.2, $w^{+++} = w^+$: in fact, given $\lambda > 0$ and $x, y \in X$,

$$w_\lambda^{+++}(x, y) = (w^+)_\lambda^{++}(x, y) = (w^+)_{\lambda+0}(x, y) = w_{\lambda+0}^+(x, y) = w_\lambda^+(x, y).$$

Similarly, $w^{---} = w^-$. Moreover, if w is continuous from the right (from the left), then $w^{++} = w$ ($w^{--} = w$, respectively).
2. The term the 'right inverse' for w^+ has been chosen from the fact that w^+ is continuous from the right. Alternatively, w^+ may be called the 'lower inverse' of w, because $w^+ \leq w^-$. Similarly, the 'left inverse' w^- of w sounds more suggestive than the 'upper inverse'.
3. The following two properties of w^-, observed in the proof of (i) and (iii) of Theorem 3.3.2, will also be useful in the sequel:

(a) if $\lambda > g^-(\mu) = w_\mu^-(x, y)$, then $g(\lambda) = w_\lambda(x, y) < \mu$;
(b) if $g(\lambda) = w_\lambda(x, y) < \mu$, then $\lambda \geq g^-(\mu) = w_\mu^-(x, y)$.

Given a nonincreasing function $g : (0, \infty) \to [0, \infty]$, its *right inverse* g^+ and *left inverse* g^- are defined by (cf. (3.3.1) and (3.3.2)): $g^+(\mu) = \inf\{\lambda > 0 : g(\lambda) \leq \mu\}$ and $g^-(\mu) = \sup\{\lambda > 0 : g(\lambda) \geq \mu\}$ for all $\mu > 0$. Clearly, functions g^+ and g^- are well-defined, map $(0, \infty)$ into $[0, \infty]$, and are nonincreasing on $(0, \infty)$. Furthermore, $g^+ \equiv 0 \Leftrightarrow g \equiv 0 \Leftrightarrow g^- \equiv 0$, and $g^+ \equiv \infty \Leftrightarrow g \equiv \infty \Leftrightarrow g^- \equiv \infty$.

Lemma 3.3.4. *Let $g : (0, \infty) \to [0, \infty]$ be a nonincreasing function. We have:*

(a) g^+ *is continuous from the right and g^- is continuous from the left on $(0, \infty)$;*
(b) $g^+(\mu) \leq g^-(\mu)$ *for all $\mu > 0$;*
(c) $g^-(\mu) = g^+(\mu - 0)$ *and $g^+(\mu) = g^-(\mu + 0)$ for all $\mu > 0$;*
(d) g *is (strictly) decreasing on $(0, \infty)$ if and only if $0 < g(\lambda) < \infty$ for all $\lambda > 0$ and $g^+(\mu) = g^-(\mu)$ for all $\mu > 0$ (and so, $g^+ = g^-$ is continuous on $(0, \infty)$);*

(e) $(g^-)^+(\lambda) = (g^+)^+(\lambda) = g(\lambda + 0) \leq g(\lambda) \leq g(\lambda - 0) = (g^-)^-(\lambda) = (g^+)^-(\lambda)$ *for all* $\lambda > 0$;

(f) $(g^+)^0 = g^0 = (g^-)^0$, *where* $g^0 = \inf\{\lambda > 0 : g(\lambda) \leq \lambda\}$ (*cf.* (2.2.1)).

Proof. (a) By (2.2.2), $g^+(\mu + 0) \leq g^+(\mu)$ and $g^-(\mu) \leq g^-(\mu - 0)$ for all $\mu > 0$. Let us establish the reverse inequalities. Suppose $g^+(\mu + 0) < \infty$, and let $\lambda > g^+(\mu + 0)$. Then, there exists $\mu_0 > \mu$ such that $\lambda > g^+(\mu')$ for all $\mu < \mu' \leq \mu_0$. By the definition of g^+, $g(\lambda) \leq \mu'$ for all $\mu < \mu' \leq \mu_0$, and so, $g(\lambda) \leq \mu$. Hence $g^+(\mu) \leq \lambda$ for all $\lambda > g^+(\mu + 0)$, which implies $g^+(\mu) \leq g^+(\mu+0)$. Thus, $g^+(\mu+0) = g^+(\mu)$, i.e., g^+ is continuous from the right at $\mu > 0$. Now, suppose $g^-(\mu) < \infty$, and let $\lambda > g^-(\mu)$. The definition of g^- implies $g(\lambda) < \mu$, and so, for each $\mu' > 0$ such that $g(\lambda) < \mu' < \mu$, we have $g^-(\mu') \leq \lambda$. Therefore, passing to the limit as $\mu' \to \mu - 0$, we get $g^-(\mu - 0) \leq \lambda$ for all $\lambda > g^-(\mu)$. This yields $g^-(\mu - 0) \leq g^-(\mu)$ and proves that $g^-(\mu - 0) = g^-(\mu)$, i.e., g^- is continuous from the left at $\mu > 0$.

(b) Assuming that $g^-(\mu) < \infty$ and $\lambda > g^-(\mu)$, we get $g(\lambda) < \mu$, which implies $g^+(\mu) \leq \lambda$, and so, as $\lambda \to g^-(\mu)$, we obtain $g^+(\mu) \leq g^-(\mu)$.

(c) By (b), $g^+(\mu') \leq g^-(\mu')$ for all $0 < \mu' < \mu$. Passing to the limit as $\mu' \to \mu - 0$, we find, by (a), $g^+(\mu - 0) \leq g^-(\mu - 0) = g^-(\mu)$. To prove the reverse inequality, we assume that $g^+(\mu - 0) < \infty$, and let $\lambda > g^+(\mu - 0)$. Then, there exists $0 < \mu_0 < \mu$ such that $\lambda > g^+(\mu')$ for all μ' with $\mu_0 \leq \mu' < \mu$, and so, by the definition of g^+, $g(\lambda) \leq \mu'$. Hence $g(\lambda) < \mu$, and the definition of g^- implies $g^-(\mu) \leq \lambda$. Taking into account the arbitrariness of $\lambda > g^+(\mu - 0)$, we get $g^-(\mu) \leq g^+(\mu - 0)$.

The second equality in (c) is established similarly.

(d)($\Rightarrow$) Since g is decreasing, given $0 < \lambda_1 < \lambda_2$, we have $0 \leq g(\lambda_2) < g(\lambda_1) \leq \infty$, i.e., $0 < g(\lambda_1)$ and $g(\lambda_2) < \infty$, and so, g is positive and finite valued. Now, by (b), it suffices to prove that $g^+(\mu) \geq g^-(\mu)$ for $\mu > 0$. This inequality will follow from the definition of g^+ if we show that, given $\lambda > 0$ with $g(\lambda) \leq \mu$, we have $\lambda \geq g^-(\mu)$. On the contrary, if $\lambda < g^-(\mu)$, then we may choose λ' such that $\lambda < \lambda' < g^-(\mu)$. Since g is decreasing, $g(\lambda) > g(\lambda')$, and the definition of g^- implies $g(\lambda') \geq \mu$, whence $g(\lambda) > \mu$, which contradicts the inequality $g(\lambda) \leq \mu$.

(d)($\Leftarrow$) Since g is nonincreasing, $g(\lambda_1) \geq g(\lambda_2)$ for all $0 < \lambda_1 < \lambda_2$. Assume that $g(\lambda_1) = g(\lambda_2) \equiv \mu$ for some $\lambda_1 < \lambda_2$. By the assumption, $0 < \mu < \infty$. The definitions of g^+ and g^- and equalities $g(\lambda_1) = \mu$ and $g(\lambda_2) = \mu$ imply $g^+(\mu) \leq \lambda_1$ and $g^-(\mu) \geq \lambda_2$. So, $\lambda_2 \leq g^-(\mu) = g^+(\mu) \leq \lambda_1$, which contradicts $\lambda_1 < \lambda_2$. Thus, $g(\lambda_1) > g(\lambda_2)$ for all $\lambda_1 < \lambda_2$, i.e., g is decreasing on $(0, \infty)$.

(e) The inequalities in the middle of the line in (e) are known from (2.2.2). In order to prove the left-hand side equalities in (e), we show that

$$(g^+)^+(\lambda) \leq (g^-)^+(\lambda) \leq g(\lambda + 0) \leq (g^+)^+(\lambda).$$

First, note that if $g_1, g_2 : (0, \infty) \to [0, \infty]$ are nonincreasing functions such that $g_1 \le g_2$ (i.e., $g_1(\lambda) \le g_2(\lambda)$ for all $\lambda > 0$), then $g_1^+ \le g_2^+$ and $g_1^- \le g_2^-$. Since $g^+ \le g^-$ (by virtue of (b)), we get $(g^+)^+ \le (g^-)^+$.

Second, we claim that $(g^-)^+(\lambda) \le g(\lambda)$ for all $\lambda > 0$. In fact, if $g(\lambda) < \infty$ and $\mu > g(\lambda)$, then, by the definition of g^-, $g^-(\mu) \le \lambda$. The definition of $(g^-)^+$ implies $(g^-)^+(\lambda) \le \mu$, and so, passing to the limit as $\mu \to g(\lambda)$, we get $(g^-)^+(\lambda) \le g(\lambda)$. Hence $(g^-)^+(\lambda') \le g(\lambda')$ for all $\lambda' > \lambda$, and letting λ' tend to $\lambda + 0$, we find, by (a), $(g^-)^+(\lambda) = (g^-)^+(\lambda + 0) \le g(\lambda + 0)$.

Third, suppose $(g^+)^+(\lambda) < \infty$. Given $\mu > (g^+)^+(\lambda)$, the definition of $(g^+)^+$ yields $g^+(\mu) \le \lambda$. If $g^+(\mu) < \lambda$, then $g(\lambda) \le \mu$, and so, $g(\lambda + 0) \le g(\lambda) \le \mu$. Now, if $g^+(\mu) = \lambda$, then, for any $\lambda' > \lambda = g^+(\mu)$, the definition of g^+ implies $g(\lambda') \le \mu$, whence, as $\lambda' \to \lambda + 0$, we get $g(\lambda + 0) \le \mu$. Since $\mu > (g^+)^+(\lambda)$ is arbitrary, the inequality $g(\lambda + 0) \le (g^+)^+(\lambda)$ follows.

The right-hand side equalities in (e) are established similarly by proving inequalities $g(\lambda - 0) \le (g^+)^-(\lambda) \le (g^-)^-(\lambda) \le g(\lambda - 0)$.

(f) 1. Let us prove the first equality $(g^+)^0 = g^0$.

   ($\le$) Suppose $g^0 < \infty$, and let $\mu > g^0$. By the definition of g^0, we have $g(\mu) \le \mu$, i.e., $\mu \in \{\lambda > 0 : g(\lambda) \le \mu\}$. So, the definition of g^+ implies $g^+(\mu) \le \mu$. It follows that $(g^+)^0 \le \mu$. Since $\mu > g^0$ is arbitrary, $(g^+)^0 \le g^0 < \infty$.

   ($\ge$) Suppose $(g^+)^0 < \infty$, and let $\lambda > (g^+)^0$. The definition of $(g^+)^0$ implies $g^+(\lambda) \le \lambda$. If $g^+(\lambda) < \lambda$, then, by the definition of g^+, $g(\lambda) \le \lambda$, and so, $g_{+0}(\lambda) = g(\lambda + 0) \le g(\lambda) \le \lambda$. If $g^+(\lambda) = \lambda$, then, for any $\lambda' > \lambda = g^+(\lambda)$, the definition of g^+ implies $g(\lambda') \le \lambda$, and so, as $\lambda' \to \lambda + 0$, we get $g_{+0}(\lambda) = g(\lambda + 0) \le \lambda$. Thus, in both cases, $g_{+0}(\lambda) \le \lambda$. Hence $(g_{+0})^0 \le \lambda$, which, together with Lemma 2.2.7, gives $g^0 = (g_{+0})^0 \le \lambda$. Passing to the limit as $\lambda \to (g^+)^0$, we find $g^0 \le (g^+)^0 < \infty$.

2. Now we establish the second equality, $g^0 = (g^-)^0$, employing the formula for g^0 from Remark 2.2.5, which is $g^0 = \sup \{\lambda > 0 : g(\lambda) \ge \lambda\}$.

   ($\le$) In view of Lemma 2.2.4(c), we may assume that $g^0 \ne 0$. Given $0 < \mu < g^0$, we have $g(\mu) \ge \mu$, and so, $\mu \in \{\lambda > 0 : g(\lambda) \ge \mu\}$. By the definition of g^-, $g^-(\mu) \ge \mu$, which implies $(g^-)^0 \ge \mu$. Since $\mu < g^0$ is arbitrary, we get $(g^-)^0 \ge g^0$, and, in particular, $(g^-)^0 \ne 0$.

   ($\ge$) Let $(g^-)^0 \ne 0$, and suppose $0 < \lambda < (g^-)^0$. Then, $g^-(\lambda) \ge \lambda$. If $g^-(\lambda) > \lambda$, then, by the definition of g^-, $g(\lambda) \ge \lambda$, and so, $g_{-0}(\lambda) = g(\lambda - 0) \ge g(\lambda) \ge \lambda$. If $g^-(\lambda) = \lambda$, then, for any $0 < \lambda' < \lambda = g^-(\lambda)$, the definition of g^- implies $g(\lambda') \ge \lambda$, and so, as $\lambda' \to \lambda - 0$, we get $g_{-0}(\lambda) = g(\lambda - 0) \ge \lambda$. Thus, in both cases, $g_{-0}(\lambda) \ge \lambda$. By Lemma 2.2.7, we get

$$g^0 = (g_{-0})^0 = \sup \{\mu > 0 : g_{-0}(\mu) \ge \mu\} \ge \lambda.$$

   Passing to the limit as $\lambda \to (g^-)^0$, we find $g^0 \ge (g^-)^0$, and, in particular, $g^0 \ne 0$.

This completes the proof of Lemma 3.3.4.   □

Remark 3.3.5. Conditions on the right in Lemma 3.3.4(d) are sharp as the following example shows. The function g cannot assume the value 0: if $g : (0, \infty) \to [0, \infty)$ is defined by $g(\lambda) = 1 - \lambda$ if $0 < \lambda \leq 1$, and $g(\lambda) = 0$ if $\lambda > 1$, then g is (only) nonincreasing and $g^+ = g^- = g$ on $(0, \infty)$. Furthermore, the function g cannot assume the value ∞: setting $g_1(\lambda) = 1/g(1/\lambda)$ for $\lambda > 0$, we find $g_1(\lambda) = \infty$ if $0 < \lambda \leq 1$, and $g_1(\lambda) = \lambda/(\lambda - 1)$ if $\lambda > 1$, and so, $g_1 : (0, \infty) \to (1, \infty]$ is nonincreasing and $g_1^+ = g_1^- = g_1$ on $(0, \infty)$.

If $a > 0$ and $g(\lambda) = a$ for all $\lambda > 0$, then (cf. also Lemma 3.3.4(b))

$$g^+(\mu) = \begin{cases} \infty & \text{if } 0 < \mu < a, \\ 0 & \text{if } \mu \geq a, \end{cases} \quad \text{and} \quad g^-(\mu) = \begin{cases} \infty & \text{if } 0 < \mu \leq a, \\ 0 & \text{if } \mu > a. \end{cases}$$

One more example: if $g(\lambda) = (1/\lambda) + 1$ for $\lambda > 0$, then (draw the graphs)

$$g^+(\mu) = g^-(\mu) = \begin{cases} \infty & \text{if } 0 < \mu \leq 1, \\ 1/(\mu - 1) & \text{if } \quad \mu > 1. \end{cases}$$

Keeping in mind that $g(\lambda) = w_\lambda(x, y)$, and in order to be specific, we study some more properties of g^+ (see Lemma 3.3.4(c)). Clearly, given $\mu > 0$, we have: $g^+(\mu) = 0 \Leftrightarrow g(+0) \leq \mu$ (or, equivalently, $g^+(\mu) > 0 \Leftrightarrow g(+0) > \mu$), and $g^+(\mu) = \infty \Leftrightarrow g(\lambda) > \mu$ for all $\lambda > 0$.

The following Lemma is a counterpart of Lemma 2.2.9.

Lemma 3.3.6. *If $g : (0, \infty) \to [0, \infty]$ is a nonincreasing function and $\lambda, \mu > 0$, then*

(a) $g^+(\mu) > \lambda$ *if and only if* $g(\lambda + 0) > \mu$;
(b) $g(\lambda - 0) < \mu$ *implies* $g^+(\mu) < \lambda$, *and* $g^+(\mu) < \lambda$ *implies* $g(\lambda - 0) \leq \mu$;
(c) $g^+(\mu) = \lambda$ *implies* $g(\lambda + 0) \leq \mu \leq g(\lambda - 0)$.

Proof.

(a)($\Rightarrow$) For any λ_1 and λ_2 such that $g^+(\mu) > \lambda_2 > \lambda_1 > \lambda$, by the definition of g^+, we have $g(\lambda_2) > \mu$, and the monotonicity of g implies $g(\lambda_1) \geq g(\lambda_2)$. Passing to the limit as $\lambda_1 \to \lambda + 0$, we get $g(\lambda + 0) \geq g(\lambda_2)$, and so, $g(\lambda + 0) > \mu$.

(a)($\Leftarrow$) If $g(\lambda + 0) > \mu$, then there exists $\lambda_0 > \lambda$ such that $g(\lambda') > \mu$ for all $\lambda < \lambda' \leq \lambda_0$. By the definition of g^+, $g^+(\mu) \geq \lambda'$ (otherwise, $g^+(\mu) < \lambda'$ implies $g(\lambda') \leq \mu$), and so, $g^+(\mu) \geq \lambda' > \lambda$.

(b) If $g(\lambda - 0) < \mu$, then there is $0 < \lambda_0 < \lambda$ such that $g(\lambda') < \mu$ for all λ' with $\lambda_0 \leq \lambda' < \lambda$. By the definition of g^+, $g^+(\mu) \leq \lambda'$, which implies $g^+(\mu) < \lambda$.

Now, suppose $g^+(\mu) < \lambda$. Given λ_1 and λ_2 such that $g^+(\mu) < \lambda_1 < \lambda_2 < \lambda$, the definition of g^+ implies $g(\lambda_1) \leq \mu$, and the monotonicity of g gives $g(\lambda_2) \leq g(\lambda_1)$. Passing to the limit as $\lambda_2 \to \lambda - 0$, we find $g(\lambda - 0) \leq g(\lambda_1)$, whence $g(\lambda - 0) \leq \mu$.

(c) is a consequence of (a) and (b).   □

Remark 3.3.7. 1. If, in Lemma 3.3.6, g is decreasing on $(0, \infty)$ (or it suffices to assume that $g(\lambda) < g(\lambda')$ for all $0 < \lambda' < \lambda$) and $g(\lambda) = \mu$, then $g^+(\mu) = \lambda$. In fact, $g(\lambda + 0) \leq g(\lambda) = \mu$ imply $g^+(\mu) \leq \lambda$ (by Lemma 3.3.6(a)), and so, $g^+(\mu) = \lambda$ (for, otherwise, if $g^+(\mu) < \lambda$ and $g^+(\mu) < \lambda' < \lambda$, then $g(\lambda') \leq \mu$, which contradicts $\mu = g(\lambda) < g(\lambda')$).

2. If $g(\lambda + 0) < g(\lambda - 0)$ and $g(\lambda + 0) \leq \mu < g(\lambda - 0)$, then $g^+(\mu) = \lambda$ (see Lemma 3.3.6(a), (b)).

3. Lemma 3.3.6 and remarks (1) and (2) above can be illustrated by functions g and g_1 from Remark 3.3.5 and, for instance, the following function:

$$\text{if } g(\lambda) = \begin{cases} 3 & \text{if } 0 < \lambda < 1, \\ 2 & \text{if } \quad \lambda = 1, \\ 1 & \text{if } \quad \lambda > 1, \end{cases} \quad \text{then } g^+(\mu) = \begin{cases} \infty & \text{if } 0 < \mu < 1, \\ 1 & \text{if } 1 \leq \mu < 3, \\ 0 & \text{if } \quad \mu \geq 3. \end{cases}$$

The following relations hold between modular spaces for w, w^+, and w^-.

Theorem 3.3.8. *Given a (pseudo)modular w on X, we have:*

(a) $X^*_{w+} = X^*_w = X^*_{w-}$;

(b) $X^0_{w+} = X^{\text{fin}}_w = X^0_{w-}$ *and* $X^{\text{fin}}_{w+} = X^0_w = X^{\text{fin}}_{w-}$;

(c) $d^0_{w+}(x, y) = d^0_w(x, y) = d^0_{w-}(x, y)$ *for all* $x, y \in X$ *(and* X^*_w*);*

(d) $d^1_{w+}(x, y) = d^1_w(x, y) = d^1_{w-}(x, y)$ *for all* $x, y \in X$ *(and* X^*_w*).*

Proof. We will prove the equalities only for w^+. Taking into account Remark 3.3.3 (3), the equalities for w^- are established similarly.

(a) To see that $X^*_{w+} \subset X^*_w$, let $x \in X^*_{w+}$. We have $w^+_\mu(x, x^\circ) < \infty$ for some $\mu > 0$. Choose $\lambda > 0$ such that $w^+_\mu(x, x^\circ) < \lambda$. By the definition of w^+, $w_\lambda(x, x^\circ) \leq \mu < \infty$, and so, $x \in X^*_w$. To show the reverse inclusion, suppose $x \in X^*_w$. Then, there exists $\lambda > 0$ such that $w_\lambda(x, x^\circ) < \infty$. Let $\mu > 0$ be such that $w_\lambda(x, x^\circ) \leq \mu$. The definition of w^+ implies $w^+_\mu(x, x^\circ) \leq \lambda < \infty$, i.e., $x \in X^*_{w+}$.

(b) If $x \in X^0_{w+}$, then $w^+_\mu(x, x^\circ) \to 0$ as $\mu \to \infty$, and so, given $\lambda > 0$, there exists $\mu_0(\lambda) > 0$ such that $w^+_\mu(x, x^\circ) < \lambda$ for all $\mu \geq \mu_0(\lambda)$ or, equivalently, $w^+_{\mu_0(\lambda)}(x, x^\circ) < \lambda$. By the definition of w^+, $w_\lambda(x, x^\circ) \leq \mu_0(\lambda) < \infty$ for all $\lambda > 0$, which implies $x \in X^{\text{fin}}_w$. Now, assume that $x \in X^{\text{fin}}_w$. Then, given $\varepsilon > 0$, $w_\varepsilon(x, x^\circ) < \infty$, and so, if $\mu_0 = \mu_0(\varepsilon) > 0$ is such that $w_\varepsilon(x, x^\circ) < \mu_0$, then $w^+_{\mu_0}(x, x^\circ) \leq \varepsilon$. Hence $w^+_\mu(x, x^\circ) \leq w^+_{\mu_0}(x, x^\circ) \leq \varepsilon$ for all $\mu \geq \mu_0$; in other words, $w^+_\mu(x, x^\circ) \to 0$ as $\mu \to \infty$, and so, $x \in X^0_{w+}$.

Exchanging w^+ and w in this proof, we get $X^0_w = X^{\text{fin}}_{w+}$.

(c) Given $x, y \in X$, Lemma 3.3.4(f), applied to the function $g(\lambda) = w_\lambda(x, y)$, $\lambda > 0$, yields $d^0_{w+}(x, y) = (g^+)^0 = g^0 = d^0_w(x, y)$.

(d) ($\leq$) Suppose $d^1_w(x, y) < \infty$, and let $\mu > d^1_w(x, y)$. Then, by (2.5.1) with $\theta = 1$, there exists $\lambda_0 = \lambda_0(\mu) > 0$ such that $\lambda_0 + w_{\lambda_0}(x, y) < \mu$. Since $w_{\lambda_0}(x, y) < \mu - \lambda_0$, the definition of w^+ implies $w^+_{\mu - \lambda_0}(x, y) \leq \lambda_0$, and so, by the definition of $d^1_{w+}, d^1_{w+}(x, y) \le (\mu - \lambda_0) + w^+_{\mu - \lambda_0}(x, y) \le \mu$. Since $\mu > d^1_w(x, y)$ is arbitrary, we get $d^1_{w+}(x, y) \le d^1_w(x, y) < \infty$. The reverse inequality ($\ge$) is proved along the same lines by exchanging w^+ and w.   □

In order to be specific, we evaluate the right inverse w^+ in the examples below.

Example 3.3.9 (modulars w of the form (1.3.1)). Here we consider modulars, already encountered in Examples 1.3.2, 2.2.2, and 2.5.5. Let (X, d) be a metric space.

1. Let $w_\lambda(x, y) = d(x, y)/\lambda^p$ with $p \ge 0$. If $p > 0$, then $w^+_\mu(x, y) = (d(x, y)/\mu)^{1/p}$ (cf. Examples 2.3.5(1) and 2.5.5(2)). In particular, if $p = 1$, then $w^+ = w$. Note that, for $p > 1$, w is convex, and its right inverse w^+ is not convex (because function $\mu \mapsto \mu w^+_\mu(x, y)$ is increasing in $\mu > 0$ if $x \ne y$). If $0 < p < 1$, then w is nonconvex and w^+ is convex (in this case $\varphi(u) = u^{1/p}$, $u \ge 0$, is a convex function).

   For $p = 0$, modular $w_\lambda(x, y) = d(x, y)$ is nonconvex, and w^+ is convex:

$$w^+_\mu(x, y) = \begin{cases} \infty & \text{if } 0 < \mu < d(x, y), \\ 0 & \text{if } \quad \mu \ge d(x, y), \end{cases} = \varphi\left(\frac{d(x, y)}{\mu}\right), \qquad (3.3.3)$$

   where $\varphi(u) = 0$ if $0 \le u \le 1$, and $\varphi(u) = \infty$ if $u > 1$ (see (1.3.5) and (1.3.9)). Modular spaces from Theorem 3.3.8 are $X^0_w = \{x^\circ\} = X^{\text{fin}}_{w+}$, $X^*_w = X = X^*_{w+}$, and $X^0_{w+} = X = X^{\text{fin}}_w$, and so, $X^0_w \subset X^0_{w+}$ properly.

   On the other hand, if we consider the left regularization of (3.3.3), that is, $w_\lambda(x, y) = \infty$ for $0 < \lambda \le d(x, y)$, and $w_\lambda(x, y) = 0$ for $\lambda > d(x, y)$, then w is a nonstrict convex modular on X, $w^+_\mu(x, y) = d(x, y)$ is strict and nonconvex, $w^{++}_\lambda(x, y) = w_{\lambda+0}(x, y)$, and $X^0_{w+} = \{x^\circ\} \subset X = X^0_w$.

2. Given $\lambda_0 > 0$, let $g(\lambda) = 1$ if $0 < \lambda < \lambda_0$, and $g(\lambda) = 0$ if $\lambda \ge \lambda_0$. The modular

$$w_\lambda(x, y) = g(\lambda)d(x, y) = \begin{cases} d(x, y) & \text{if } 0 < \lambda < \lambda_0, \\ 0 & \text{if } \quad \lambda \ge \lambda_0, \end{cases}$$

   is nonstrict and nonconvex, and its right inverse is given by

$$w^+_\mu(x, y) = \begin{cases} \lambda_0 & \text{if } 0 < \mu < d(x, y), \\ 0 & \text{if } \quad \mu \ge d(x, y), \end{cases} = \varphi\left(\frac{d(x, y)}{\mu}\right),$$

   where $\varphi(u) = 0$ if $0 \le u \le 1$, and $\varphi(u) = \lambda_0$ if $u > 1$. So, w^+ is nonstrict and nonconvex. (Formally, w^+ is the same as (3.3.3) for $\lambda_0 = \infty$.)

3. Consider the following nonstrict (and convex for $p \ge 1$) modular

$$w_\lambda(x, y) = \frac{d(x, y)}{\lambda^p} \text{ if } 0 < \lambda < 1, \quad w_\lambda(x, y) = 0 \text{ if } \lambda \ge 1 \quad (p > 0).$$

Its right inverse w^+ is of the form, which is strict and nonconvex:

$$w_\mu^+(x, y) = 1 \text{ if } 0 < \mu < d(x, y), \quad w_\mu^+(x, y) = \left(\frac{d(x, y)}{\mu}\right)^{1/p} \text{ if } \mu \geq d(x, y)$$

(the nonconvexity of w^+ follows from Proposition 1.2.3(b)). Moreover,

$$d_w^0(x, y) = d_{w^+}^0(x, y) = \min\{1, (d(x, y))^{1/(p+1)}\} \quad (p > 0),$$
$$d_w^*(x, y) = \min\{1, (d(x, y))^{1/p}\} \quad (p \geq 1). \tag{3.3.4}$$

4. For $g(\lambda) = \max\{1, 1/\lambda^p\}$ $(p > 0)$, we have a strict nonconvex modular

$$w_\lambda(x, y) = \frac{d(x, y)}{\lambda^p} \text{ if } 0 < \lambda < 1, \quad w_\lambda(x, y) = d(x, y) \text{ if } \lambda \geq 1,$$

and so, its right inverse w^+ is of the form

$$w_\mu^+(x, y) = \infty \text{ if } 0 < \mu < d(x, y), \quad w_\mu^+(x, y) = \left(\frac{d(x, y)}{\mu}\right)^{1/p} \text{ if } \mu \geq d(x, y),$$

which is strict, and convex for $0 < p \leq 1$. Also, for modular spaces, we have the following: $X_w^0 = \{x^\circ\}, X_{w^+}^0 = X_{w^+}^* = X_w^* = X$, and

$$d_w^0(x, y) = d_{w^+}^0(x, y) = \max\{d(x, y), (d(x, y))^{1/(p+1)}\}.$$

5. If $w_\lambda(x, y) = d(x, y)/(h(\lambda) + d(x, y))$ is the modular from (1.3.2) (and Example 2.2.12), then

$$w_\mu^+(x, y) = \left(\left(\frac{1}{\mu} - 1\right)d(x, y)\right)^{1/p} \text{ if } 0 < \mu < 1, \quad \text{and} = 0 \text{ if } \mu \geq 1.$$

Example 3.3.10 (modulars w of the form (1.3.5)). Suppose $\varphi : [0, \infty) \to [0, \infty]$ is a nondecreasing function such that $\varphi(0) = 0$ and $\varphi \neq 0$. The function $\varphi_+^{-1} :$ $[0, \infty) \to [0, \infty]$, defined by $\varphi_+^{-1}(v) = \sup\{u \geq 0 : \varphi(u) \leq v\}$ for all $v \geq 0$, is called the *right inverse* of φ. Clearly, φ_+^{-1} is nondecreasing and continuous from the right on $[0, \infty)$.

1. If $w_\lambda(x, y) = \varphi(d(x, y)/\lambda)$, $\lambda > 0$, $x, y \in X$, is a modular on X of the form (1.3.5), then (under conventions $1/0 = \infty$, $1/\infty = 0$, and $\infty \cdot 0 = 0$)

$$w_\mu^+(x, y) = \frac{1}{\varphi_+^{-1}(\mu)} \cdot d(x, y) \quad \text{for all} \quad \mu > 0 \quad \text{and} \quad x, y \in X. \tag{3.3.5}$$

This equality means that the right inverse of a modular from (1.3.5) is a modular of the form (1.3.1) (and vice versa).

To prove (3.3.5), we assume that $x \neq y$, and set $a = d(x, y) > 0$. Since

$$w_\mu^+(x, y) = \inf\{\lambda > 0 : \varphi(a/\lambda) \leq \mu\} = a \cdot \inf\{1/u : u > 0 \text{ and } \varphi(u) \leq \mu\},$$

it suffices to show that

$$\inf\{1/u > 0 : \varphi(u) \leq \mu\} = 1/\varphi_+^{-1}(\mu) \quad \text{for all} \quad \mu > 0. \tag{3.3.6}$$

On the one hand, since $\varphi_+^{-1}(\mu) = \infty$ is equivalent to $\{u > 0 : \varphi(u) \leq \mu\} = (0, \infty)$, which is equivalent to $\{1/u > 0 : \varphi(u) \leq \mu\} = (0, \infty)$, equality (3.3.6) follows. On the other hand, $\varphi_+^{-1}(\mu) = 0$ is equivalent to $\{u > 0 : \varphi(u) > \mu\} = (0, \infty)$, which is equivalent to $\{1/u > 0 : \varphi(u) \leq \mu\} = \varnothing$, and, once again, (3.3.6) follows. Now, suppose $0 < \varphi_+^{-1}(\mu) < \infty$, so that the left-hand side in (3.3.6) is also positive and finite. If $u > 0$ is such that $\varphi(u) \leq \mu$, then, by the definition of φ_+^{-1}, $u \leq \varphi_+^{-1}(\mu)$, and so, $1/u \geq 1/\varphi_+^{-1}(\mu)$. This establishes the inequality $(\geq)$ in (3.3.6). To prove the reverse inequality, let $\xi > 1/\varphi_+^{-1}(\mu)$. Since $1/\xi < \varphi_+^{-1}(\mu)$, the definition of φ_+^{-1} implies $\varphi(1/\xi) \leq \mu$, and so (putting $u = 1/\xi$), we get

$$\{\xi > 0 : \xi > 1/\varphi_+^{-1}(\mu)\} \subset \{1/u > 0 : \varphi(u) \leq \mu\}.$$

Taking the infima, we obtain the inequality $(\leq)$ in (3.3.6).

2. Here we assume that $\varphi : [0, \infty) \to [0, \infty)$ is a nondecreasing and unbounded function such that $\varphi(+0) = 0$, and $\varphi(u) > 0$ for $u > 0$. By virtue of Theorem 3.3.8(c) and (3.3.5), given $x, y \in X$, we have

$$d_w^0(x, y) = d_{w+}^0(x, y) = \inf\{\mu > 0 : w_\mu^+(x, y) \leq \mu\}$$
$$= \inf\{\mu > 0 : \mu\varphi_+^{-1}(\mu) \geq d(x, y)\} = \Phi_-^{-1}(d(x, y)),$$

where the function $\Phi : [0, \infty) \to [0, \infty)$, given by $\Phi(u) = u\varphi_+^{-1}(u)$ for $u > 0$ and $\Phi(0) = 0$, is increasing on $[0, \infty)$, and $\Phi_-^{-1}(v) = \inf\{u \geq 0 : \Phi(u) \geq v\}$, $v \geq 0$, is the *left inverse* of Φ, which is nondecreasing and continuous on $[0, \infty)$.

For instance, let $\varphi(u) = u$ if $0 \leq u \leq 1$, $\varphi(u) = 1$ if $1 < u < 2$, and $\varphi(u) = u - 1$ if $u \geq 2$. Then $\varphi_+^{-1}(\mu) = \mu$ if $0 \leq \mu < 1$, and $\varphi_+^{-1}(\mu) = \mu + 1$ if $\mu \geq 1$, and so, $\Phi_-^{-1}(v) = \sqrt{v}$ if $0 \leq v \leq 1$, $\Phi_-^{-1}(v) = 1$ if $1 < v < 2$, and $\Phi_-^{-1}(v) = (\sqrt{4v + 1} - 1)/2$ if $v \geq 2$ (cf. Example 2.3.5(6)).

3. Suppose $\varphi(u) = 0$ if $0 \leq u \leq 1$, and $\varphi(u) = \log u$ if $u > 1$. Then

$$w_\lambda(x, y) = \varphi\left(\frac{d(x, y)}{\lambda}\right) = \begin{cases} \log(d(x, y)/\lambda) & \text{if } 0 < \lambda < d(x, y), \\ 0 & \text{if } \quad \lambda \geq d(x, y). \end{cases}$$

Since $\varphi_+^{-1}(v) = e^v$, $v \geq 0$, formula (3.3.5) gives $w_\mu^+(x, y) = e^{-\mu}d(x, y)$, which is the modular from Example 2.5.5(4).

4. Consider the modular w from Example 2.3.5(5). We have

$$
w_\mu^+(x, y) = \begin{cases} 0 & \text{if } x = y \text{ and } \mu > 0, \\ \infty & \text{if } x \neq y \text{ and } 0 < \mu < 1, \\ d(x, y)/\mu & \text{if } x \neq y \text{ and } \mu \geq 1. \end{cases} \tag{3.3.7}
$$

This modular is strict and convex on X, and so,

$$
d_{w+}^*(x, y) = \inf\{\mu > 0 : w_\mu^+(x, y) \leq 1\} = \begin{cases} 0 & \text{if } x = y, \\ \max\{1, d(x, y)\} & \text{if } x \neq y. \end{cases} \tag{3.3.8}
$$

5. Let $w_\lambda(x, y) = \sup_{n \in \mathbb{N}} \big(d(x_n, y_n)/\lambda\big)^{1/n}$ be the modular from (2.4.1) with $\lambda > 0$ and $x = \{x_n\}, y = \{y_n\} \in X = M^{\mathbb{N}}$, where (M, d) is a metric space. The right inverse modular w^+ is given, for $\mu > 0$ and $x, y \in X$, by

$$
w_\mu^+(x, y) = \sup_{n \in \mathbb{N}} \frac{d(x_n, y_n)}{\mu^n}. \tag{3.3.9}
$$

Clearly, w^+ is strict and, by Theorem 3.3.2, continuous from the right. (It is not continuous from the left: e.g., if $M = \mathbb{R}$, $x, y \in \mathbb{R}$, $x \neq y$, $x = \{x\}_{n=1}^\infty$ and $y = \{y\}_{n=1}^\infty$, then $w_\mu^+(x, y) = \infty$ for all $0 < \mu < 1$, and $w_1(x, y) = |x - y|$.) Moreover, w^+ is convex: in fact, since, for each $n \in \mathbb{N}$, the function $\lambda \mapsto \lambda/\lambda^n$ is nonincreasing on $(0, \infty)$, we have

$$
\frac{1}{(\lambda + \mu)^n} \leq \frac{\lambda}{\lambda + \mu} \cdot \frac{1}{\lambda^n} \quad \text{for all} \quad \lambda, \mu > 0,
$$

and so,

$$
\frac{d(x_n, y_n)}{(\lambda + \mu)^n} \leq \frac{d(x_n, z_n) + d(z_n, y_n)}{(\lambda + \mu)^n} \leq \frac{\lambda}{\lambda + \mu} \cdot \frac{d(x_n, z_n)}{\lambda^n} + \frac{\mu}{\lambda + \mu} \cdot \frac{d(z_n, y_n)}{\mu^n}
$$

$$
\leq \frac{\lambda}{\lambda + \mu} w_\lambda^+(x, z) + \frac{\mu}{\lambda + \mu} w_\mu^+(z, y) \quad \text{for all} \quad n \in \mathbb{N}.
$$

Taking the supremum over all $n \in \mathbb{N}$, we obtain the convexity property (axiom (iv)). Hence (Sect. 2.1) $X_{w+}^0 = X_{w+}^*$, and metric d_{w+}^* from (2.3.3) is well-defined on X_{w+}^*:

$$
d_{w+}^*(x, y) = \sup_{n \in \mathbb{N}} \big(d(x_n, y_n)\big)^{1/n} = w_1(x, y). \tag{3.3.10}
$$

By virtue of Theorem 3.3.8 (see also Sect. 2.4.2), we have

$$
X_{w+}^{\text{fin}} = X_w^0 \subset X_w^* = X_{w+}^* = X_{w+}^0 = X_w^{\text{fin}}. \tag{3.3.11}
$$

The first two equalities provide alternative descriptions of modular spaces X_w^0 and X_w^* (see Proposition 2.4.2 and (2.4.3)) as follows:

$$x \in X_w^0(x^\circ) \Leftrightarrow \{\lambda^n d(x_n, x_n^\circ)\}_{n=1}^\infty \text{ is bounded in } \mathbb{R} \text{ for } all \ \lambda > 0;$$

$$x \in X_w^*(x^\circ) \Leftrightarrow \{\lambda^n d(x_n, x_n^\circ)\}_{n=1}^\infty \text{ is bounded in } \mathbb{R} \text{ for } some \ \lambda > 0.$$

Example 3.3.11 (the right inverse of the Hausdorff pseudomodular). Let w be a (pseudo)modular on X. Denote the quantities (1.3.12), (1.3.13) and (1.3.14) by $E_\lambda^{(w)}(A, B)$, showing the dependence on w explicitly. By virtue of Sect. 1.3.5,

$$W_\lambda^{(w)}(A, B) = \max\{E_\lambda^{(w)}(A, B), E_\lambda^{(w)}(B, A)\}, \quad A, B \subset X,$$

is the Hausdorff pseudomodular on $\mathscr{P}(X)$, induced by w. Let w^+ be the right inverse of w and $W^{(w)+}$ be the right inverse of $W^{(w)}$. Given $A, B \in \mathscr{P}(X)$, the following inequalities hold:

$$W_\lambda^{(w^+)}(A, B) \leq W_\mu^{(w)+}(A, B) \leq W_\mu^{(w^+)}(A, B) \quad \text{for all} \quad 0 < \mu < \lambda.$$

The proof follows the lines of the proof of Theorem 2.2.13 with suitable modifications, and so, is omitted.

3.4   Convex Right Inverses

As we have seen in Examples 3.3.9 and 3.3.10, the right w^+ (and left w^-) inverse of a (convex) modular w may be nonconvex. Now, we exhibit a transform of (pseudo)modulars, which preserves the convexity property of w and the two metrics d_w^* from (2.3.3) and d_w^{1*} from Theorem 2.5.7 (with $\theta = 1$).

Recall that if w is a convex (pseudo)modular on X, then the function $\hat{w}_\lambda(x, y) \equiv (w^\wedge)_\lambda(x, y) = \lambda w_\lambda(x, y)$ is a (pseudo)modular on X. Given $\mu > 0$ and $x, y \in X$, we define the *convex right inverse* of w by

$$w_\mu^{(+c)}(x, y) = \frac{\hat{w}_\mu^+(x, y)}{\mu}, \quad \text{where} \quad \hat{w}^+ = (w^\wedge)^+.$$

By virtue of Sect. 3.2 and Theorem 3.3.2, $w^{(+c)}$ is a convex (pseudo)modular on X, which is continuous from the right on $(0, \infty)$.

Theorem 3.4.1. *Let w be a convex (pseudo)modular on X. Given $x, y \in X$, we have:*

(a) $w_\mu^{(+c)}(x, y) = \inf\{\lambda > 0 : \lambda w_{\lambda\mu}(x, y) \leq 1\}$ *for all* $\mu > 0$;

(b) $w_\lambda^{(+c)(+c)}(x, y) = w_{\lambda+0}(x, y) \leq w_\lambda(x, y)$ *for all* $\lambda > 0$;

(c) $d_{w^{(+c)}}^*(x, y) = d_w^*(x, y)$ *and* $d_{w^{(+c)}}^{1*}(x, y) = d_w^{1*}(x, y)$;

(d) $X_w^{\mathrm{fin}} \subset X_{w(+c)}^0 = X_{w(+c)}^* = X_w^* = X_w^0 \supset X_{w(+c)}^{\mathrm{fin}}$.

Proof. (a) Changing the variables in the infimum, we get

$$w_\mu^{(+c)}(x,y) = \frac{1}{\mu}\inf\{\lambda > 0 : \hat{w}_\lambda(x,y) \le \mu\} = \inf\{\lambda/\mu > 0 : \lambda w_\lambda(x,y) \le \mu\}$$

$$= \inf\{\lambda > 0 : \lambda\mu w_{\lambda\mu}(x,y) \le \mu\} = \inf\{\lambda > 0 : \lambda w_{\lambda\mu}(x,y) \le 1\}.$$

(b) Since $\widehat{w_\mu^{(+c)}}(x,y) = \mu w_\mu^{(+c)}(x,y) = \hat{w}_\mu^+(x,y)$, $\mu > 0$, taking into account Theorem 3.3.2, we find

$$w_\lambda^{(+c)(+c)}(x,y) = \frac{1}{\lambda}\left(\widehat{w^{(+c)}}\right)_\lambda^+(x,y) = \frac{1}{\lambda}\inf\{\mu > 0 : \widehat{w_\mu^{(+c)}}(x,y) \le \lambda\}$$

$$= \frac{1}{\lambda}\hat{w}_\lambda^{++}(x,y) = \frac{1}{\lambda}\hat{w}_{\lambda+0}(x,y) = \frac{1}{\lambda}\cdot\lambda w_{\lambda+0}(x,y).$$

(c) By definition (2.3.3) and Theorem 3.3.8(c), we have

$$d_{w(+c)}^*(x,y) = \inf\{\mu > 0 : w_\mu^{(+c)}(x,y) \le 1\} = \inf\{\mu > 0 : \hat{w}_\mu^+(x,y) \le \mu\}$$

$$= d_{\hat{w}+}^0(x,y) = d_{\hat{w}}^0(x,y) = \inf\{\lambda > 0 : \hat{w}_\lambda(x,y) \le \lambda\}$$

$$= \inf\{\lambda > 0 : w_\lambda(x,y) \le 1\} = d_w^*(x,y).$$

The second equality in (c) follows from (2.5.1) and Theorem 3.3.8(d):

$$d_{w(+c)}^{1*}(x,y) = \inf_{\lambda>0}\left(\lambda + \lambda w_\lambda^{(+c)}(x,y)\right) = \inf_{\lambda>0}\left(\lambda + \hat{w}_\lambda^+(x,y)\right)$$

$$= d_{\hat{w}+}^1(x,y) = d_{\hat{w}}^1(x,y) = \inf_{\lambda>0}\left(\lambda + \hat{w}_\lambda(x,y)\right)$$

$$= \inf_{\lambda>0}\left(\lambda + \lambda w_\lambda(x,y)\right) = d_w^{1*}(x,y).$$

(d) The first inclusion $\subset$ follows from (2.3.1) and Theorem 3.3.8(b):

$$X_w^{\mathrm{fin}} = X_{\hat{w}}^{\mathrm{fin}} = X_{\hat{w}+}^0 \subset X_{w(+c)}^0,$$

and the last inclusion $\supset$ is a consequence of Theorem 3.3.8(b) and (2.3.1):

$$X_{w(+c)}^{\mathrm{fin}} = X_{\hat{w}+}^{\mathrm{fin}} = X_{\hat{w}}^0 \subset X_w^0.$$

Since $w^{(+c)}$ and w are convex, we get the first and last equalities in (d). Finally, by Theorem 3.3.8(a) and (2.3.1), we obtain the equality in the middle:

$$X_{w(+c)}^* = X_{\hat{w}+}^* = X_{\hat{w}}^* = X_w^*. \qquad \square$$

Example 3.4.2 (modulars $w^{(+c)}$).

1. Let $w_\lambda(x, y) = d(x, y)/\lambda^p$ $(p \geq 1)$ be the strict convex modular from Example 3.3.9(1). By Theorem 3.4.1(a), we have: if $p > 1$, then

$$w_\mu^{(+c)}(x, y) = \left(\frac{d(x, y)}{\mu^p}\right)^{1/(p-1)} = \frac{1}{\mu} \cdot \left(\frac{d(x, y)}{\mu}\right)^{1/(p-1)},$$

and if $p = 1$, then

$$w_\mu^{(+c)}(x, y) = \begin{cases} \infty & \text{if } 0 < \mu < d(x, y), \\ 0 & \text{if } \quad \mu \geq d(x, y). \end{cases}$$

Furthermore, all modular spaces from Theorem 3.4.1(d) are equal to X, except that in the case $p = 1$ we have $X_{w^{(+c)}}^{\text{fin}}(x^\circ) = \{x^\circ\}$.

For nonconvex modulars w, e.g., when $0 \leq p < 1$ above, its convex right inverse $w^{(+c)}$ may degenerate: $w_\mu^{(+c)}(x, y) = \inf\{\lambda > 0 : \lambda^{1-p}d(x, y) \leq 1\} \equiv 0$.

2. For the convex modular $w_\lambda(x, y) = d(x, y)/\lambda^p$ if $0 < \lambda < 1$ $(p \geq 1)$, and $w_\lambda(x, y) = 0$ if $\lambda \geq 1$, from Example 3.3.9(3), we have

$$w_\mu^{(+c)}(x, y) = \begin{cases} 1/\mu & \text{if } 0 < \mu < d(x, y), \\ \left(d(x, y)/\mu^p\right)^{1/(p-1)} & \text{if } \quad \mu \geq d(x, y), \end{cases} \quad \text{if } p > 1,$$

and

$$w_\mu^{(+c)}(x, y) = \begin{cases} 1/\mu & \text{if } 0 < \mu < d(x, y), \\ 0 & \text{if } \quad \mu \geq d(x, y), \end{cases} \quad \text{if } p = 1.$$

The value $d_{w^{(+c)}}^*(x, y)$ is given by equality (3.3.4) (cf. also Theorem 3.4.1(c)).

3. Let $g(\lambda) = \infty$ for $0 < \lambda < \lambda_0$, and $g(\lambda) = 0$ for $\lambda \geq \lambda_0$. Define w by $w_\lambda(x, y) = g(\lambda)\delta(x, y)$ (i.e., $= \infty$ for $x \neq y$ and $0 < \lambda < \lambda_0$, and $= 0$ otherwise). We have $w_\mu^{(+c)}(x, y) = (\lambda_0/\mu)\delta(x, y)$.

4. Consider modular (3.3.7) from Example 3.3.10(4), written, for $x \neq y$, as

$$w_\lambda(x, y) = \begin{cases} \infty & \text{if } 0 < \lambda < 1, \\ d(x, y)/\lambda & \text{if } \quad \lambda \geq 1, \end{cases} \quad (\text{and } = 0 \text{ if } x = y \text{ and } \lambda > 0).$$

Then, for $x \neq y$, we have

$$w_\mu^{(+c)}(x, y) = \begin{cases} \infty & \text{if } 0 < \mu < d(x, y), \\ 1/\mu & \text{if } \quad \mu \geq d(x, y), \end{cases} \quad (\text{and } = 0 \text{ if } x = y \text{ and } \mu > 0),$$

and metric $d_{w^{(+c)}}^*(x, y)$ is given by the right-hand side of (3.3.8).

5. Rewriting modular (3.3.9) as $w_\lambda(x, y) = \sup_{n\in\mathbb{N}} d(x_n, y_n)/\lambda^n$, we get, by virtue of Theorem 3.4.1(a),

$$
w_\mu^{(+c)}(x, y) = \begin{cases} \infty & \text{if } 0 < \mu < d(x_1, y_1), \\ \sup_{n\geq 2}\left(\dfrac{d(x_n, y_n)}{\mu^n}\right)^{1/(n-1)} & \text{if } \quad \mu \geq d(x_1, y_1), \end{cases}
$$

and $d_{w^{(+c)}}^*(x, y)$ is expressed by the middle term of (3.3.10).

Remark 3.4.3. 1. Any modular w on X can be transformed into a convex modular w' as follows:

$$
w'_\mu(x, y) = \frac{w_\mu^+(x, y)}{\mu} = \inf\{\lambda > 0 : w_{\lambda\mu}(x, y) \leq \mu\},
$$

in which case we have $d_{w'}^*(x, y) = d_w^0(x, y)$.
2. Given a convex modular w on X, we may set

$$
w''_\mu(x, y) = \hat{w}_\mu^+(x, y) = \inf\{\lambda > 0 : \lambda w_\lambda(x, y) \leq \mu\},
$$

and so, $d_{w''}^0(x, y) = d_{\hat{w}+}^0(x, y) = d_{\hat{w}}^0(x, y) = d_w^*(x, y)$.

3.5 Bibliographical Notes and Comments

A classical modular ρ on a linear space X generates the family $\{\rho_\varepsilon\}_{\varepsilon>0}$ according to the rule $\rho_\varepsilon(x) = \rho(x/\varepsilon)$ for all $x \in X$ (e.g., $|x|_\rho = \inf\{\varepsilon > 0 : \rho(x/\varepsilon) \leq \varepsilon\}$). It follows that any family of the form $\{(1/\varepsilon^p)\rho(x/\varepsilon)\}_{\varepsilon>0}$ is no longer a classical modular ($p > 0$ is fixed). Hence multiplication of a classical modular by a non-increasing function, as in (1.3.1), leads outside the scope of classical modulars. Metric modulars $w = \{w_\lambda(x, y)\}_{\lambda>0}$ lack this deficiency and are more flexible and transformable.

Section 3.1. The material of Proposition 3.1.1 and Sect. 3.1.1 is taken from Chistyakov [23]. In Sect. 3.1.1, it is shown that the notion of φ-convex modulars generalizes the notion of classical s-convex modulars ($0 < s \leq 1$) due to Orlicz [90] if we set $\varphi(u) = u^{1/s}$ for $u \geq 0$. Inequalities (3.1.1) are proved in [23, Theorem 6]. Definition 3.1.3 is new. In particular, it gives, in Sect. 3.1.3, a generalization of classical complex modulars considered in Musielak [75] and Rolewicz [95]. Various axioms are known in the literature to define modulars on linear spaces or spaces with a prescribed algebraic structure: Leśniewicz [60, 61], Leśniewicz and Orlicz [63], Musielak [74, 75], Musielak and Peetre [79], Nakano [80–82], Nowak [84, 85], Orlicz [89], Turpin [102], Yamamuro [106].

Section 3.2. We present modular transforms resembling metric transforms, see Deza and Deza [36, Sect. 4.1].

Sections 3.3 and 3.4. The inverses of a metric (pseudo)modular are a powerful tool to generate new (pseudo)modulars from given ones. Definition of w^+ and its initial properties are presented in Chistyakov [24, Theorem 2.17]. A thorough study of inverses is given in the sections under consideration.

Chapter 4
Topologies on Modular Spaces

Abstract In this chapter, we introduce, study and compare two kinds of convergences and topologies, induced by a pseudomodular on a set—metric and modular.

4.1 The Metric Convergence and Topology

Throughout this section, w is a (pseudo)modular on a set X. Here we study *metric notions* in the extended (pseudo)metric space (X, d_w^0) and (pseudo)metric spaces (X_w^*, d_w^0), and (X_w^*, d_w^*) when w is convex.

4.1.1 The Metric Convergence

We begin by considering the d_w^0-convergence of sequences from X, also called (by some abuse of terminology) the *metric convergence*: a sequence $\{x_n\} \subset X$ *converges* to $x \in X$ if $d_w^0(x_n, x) \to 0$, which is denoted simply by $x_n \to x$ (if it does not lead to ambiguities, treating a convergence of sequences, we usually omit 'as $n \to \infty$').

Our primary aim is to characterize this convergence in terms of w.

A set $A \subset X$ is said to be *closed with respect to the metric convergence* if, given a sequence $\{x_n\} \subset A$ and $x \in X$ such that $x_n \to x$, we have $x \in A$.

Theorem 4.1.1. (a) *Given a sequence $\{x_n\} \subset X$ and $x \in X$, we have:*

$$x_n \to x \ \textit{ if and only if } \ w_\lambda(x_n, x) \to 0 \textit{ for all } \lambda > 0. \qquad (4.1.1)$$

Moreover, if w is convex, then (4.1.1) is also equivalent to $d_w^*(x_n, x) \to 0$, and if w is a modular on X, then the limit x from (4.1.1) is unique.

(b) *An assertion similar to (4.1.1) holds for Cauchy sequences.*

(c) *The modular spaces X_w^*, X_w^0, and X_w^{fin} (around $x^\circ \in X$) are closed with respect to the metric convergence.*

© Springer International Publishing Switzerland 2015
V.V. Chistyakov, *Metric Modular Spaces*, SpringerBriefs in Mathematics,
DOI 10.1007/978-3-319-25283-4_4

Proof. (a) Let $x_n \to x$. Given $\varepsilon > 0$, there exists $n_0 = n_0(\varepsilon) \in \mathbb{N}$ such that $d_w^0(x_n, x) < \varepsilon$ for all $n \geq n_0(\varepsilon)$. By (1.2.4) and Theorem 2.2.11(a),

$$w_\varepsilon(x_n, x) \leq w_{\varepsilon-0}(x_n, x) < \varepsilon \quad \text{for all} \quad n \geq n_0(\varepsilon). \qquad (4.1.2)$$

Hence, if $\lambda > 0$, then, for any $n \geq n_0(\min\{\varepsilon, \lambda\})$, we find, by (1.2.1),

$$w_\lambda(x_n, x) \leq w_{\min\{\varepsilon,\lambda\}}(x_n, x) < \min\{\varepsilon, \lambda\} \leq \varepsilon,$$

and so, $w_\lambda(x_n, x) \to 0$. Conversely, if $\varepsilon > 0$, then $w_\varepsilon(x_n, x) \to 0$, and so, there is $n_1(\varepsilon) \in \mathbb{N}$ such that $w_\varepsilon(x_n, x) \leq \varepsilon$ for all $n \geq n_1(\varepsilon)$. The definition of d_w^0 implies $d_w^0(x_n, x) \leq \varepsilon$ for $n \geq n_1(\varepsilon)$, i.e., $d_w^0(x_n, x) \to 0$.

If w is convex, the equivalence of convergences $d_w^*(x_n, x) \to 0$ and $d_w^0(x_n, x) \to 0$ is a consequence of Theorem 2.3.1 (see (2.3.6)).

If w is a modular on X, then, by Theorem 2.2.1 (or Theorem 2.3.1), d_w^0 (or d_w^*) is an extended metric on X and a metric on X_w^*, and so, the limit x from (4.1.1) is uniquely determined.

(b) The assertion for Cauchy sequences $\{x_n\}$ is of the form:

$$\lim_{n,m\to\infty} d_w^0(x_n, x_m) = 0 \quad \text{if and only if} \quad \lim_{n,m\to\infty} w_\lambda(x_n, x_m) = 0 \quad \text{for all} \quad \lambda > 0,$$

and its proof is similar to the one given for (4.1.1).

(c) Suppose $\{x_n\} \subset X$, $x \in X$, and $x_n \to x$. Given $\varepsilon > 0$, by virtue of (4.1.2) and axiom (iii), if $\lambda_0 > 0$, then we have, for $n_0 = n_0(\varepsilon)$,

$$w_{\varepsilon+\lambda_0}(x, x^\circ) \leq w_\varepsilon(x, x_{n_0}) + w_{\lambda_0}(x_{n_0}, x^\circ) \leq \varepsilon + w_{\lambda_0}(x_{n_0}, x^\circ). \qquad (4.1.3)$$

If $\{x_n\} \subset X_w^*$, then $x_{n_0} \in X_w^*$, and so, there is $\lambda_0 > 0$ such that $w_{\lambda_0}(x_{n_0}, x^\circ) < \infty$. By (4.1.3), $w_{\varepsilon+\lambda_0}(x, x^\circ) < \infty$, which means that $x \in X_w^*$.

Suppose $\{x_n\} \subset X_w^0$. Then, $x_{n_0} \in X_w^0$ implies $w_\lambda(x_{n_0}, x^\circ) \to 0$ as $\lambda \to \infty$, and so, $w_{\lambda_0}(x_{n_0}, x^\circ) < \varepsilon$ for some $\lambda_0 = \lambda_0(\varepsilon) > 0$. By (1.2.1) and (4.1.3), we find $w_\lambda(x, x^\circ) \leq w_{\varepsilon+\lambda_0}(x, x^\circ) < 2\varepsilon$ for all $\lambda \geq \varepsilon + \lambda_0$, which implies $w_\lambda(x, x^\circ) \to 0$ as $\lambda \to \infty$, i.e., $x \in X_w^0$.

Finally, let $\{x_n\} \subset X_w^{\text{fin}}$, and $\lambda > 0$ be arbitrary. Setting $\varepsilon = \lambda/2$ and $\lambda_0 = \lambda/2$ in (4.1.2) and (4.1.3), we get, from $x_{n_0} \in X_w^{\text{fin}}$, that $w_{\lambda_0}(x_{n_0}, x^\circ) < \infty$, and so, (4.1.3) implies $w_\lambda(x, x^\circ) = w_{\varepsilon+\lambda_0}(x, x^\circ) < \infty$. Thus, $x \in X_w^{\text{fin}}$. $\qquad \square$

4.1.2 The Metric Topology

1. *Open sets.* Given $x \in X$ and $r > 0$, the *open ball* of radius r centered at x is the set

$$B(x, r) \equiv B^{d_w^0}(x, r) = \{y \in X : d_w^0(x, y) < r\}.$$

A nonempty set $U \subset X$ is *open* if for every $x \in U$ there exists $r = r(x) > 0$ such that $B(x, r) \subset U$. Note that, by Theorem 2.2.1, if $x \in X_w^*(x^\circ)$, then $B(x, r) \subset X_w^*(x^\circ)$ for all $r > 0$. As a consequence, $U \subset X$ is open if and only if $U \cap X_w^*(x^\circ)$ is open for all $x^\circ \in X$. Thus, we may restrict our considerations to subsets of the modular space X_w^* (with an arbitrary fixed center $x^\circ \in X$). An open ball in X_w^*, the empty set, and X_w^* itself are examples of open sets. The *interior* of a set $A \subset X_w^*$, denoted by A°, is the largest open set contained in A (i.e., the union of all open sets $U \subset A$). Clearly, $A^\circ = \{x \in A : B(x, r) \subset A \text{ for some } r > 0\}$, $A^\circ \subset A$, and A is open if and only if $A^\circ = A$.

We denote by $\tau(d_w^0)$ the *metric topology* on X_w^*, induced by d_w^0 (i.e., the family of all open subsets of X_w^*). Clearly, $U \in \tau(d_w^0)$ if and only if its complement $U^c = X_w^* \setminus U$ is closed with respect to the metric convergence.

2. *Modular entourages.* Given $\lambda, \mu > 0$ and $x \in X_w^*$, the *w-entourage* (or *modular entourage*) about x relative to λ and μ is the set

$$B_{\lambda,\mu}(x) \equiv B_{\lambda,\mu}^w(x) = \{y \in X_w^* : w_\lambda(x, y) < \mu\}.$$

The entourages for the right w_{+0} and left w_{-0} regularizations of w (cf. (1.2.2) and (1.2.3)) will be denoted by $B_{\lambda+0,\mu}(x)$ and $B_{\lambda-0,\mu}(x)$, respectively.

Clearly, $x \in B_{\lambda,\mu}(x)$, and $B_{\lambda,\mu_1}(x) \subset B_{\lambda,\mu_2}(x)$ if $0 < \mu_1 < \mu_2$. By virtue of inequalities (1.2.4), given $0 < \lambda_1 < \lambda < \lambda_2$ and $\mu > 0$, we have

$$B_{\lambda_1+0,\mu}(x) \subset B_{\lambda-0,\mu}(x) \subset B_{\lambda,\mu}(x) \subset B_{\lambda+0,\mu}(x) \subset B_{\lambda_2-0,\mu}(x),$$

and so, mappings $\lambda \mapsto B_{\lambda,\mu}(x)$ and $\mu \mapsto B_{\lambda,\mu}(x)$ are *nondecreasing* (in the sense of the inclusion relation). By Theorem 2.2.11(a), $B(x, \lambda) = B_{\lambda-0,\lambda}(x)$ for $\lambda > 0$. Hence

$$B_{\mu,\mu}(x) \subset B_{\mu,\lambda}(x) \subset B(x, \lambda) = B_{\lambda-0,\lambda}(x) \subset B_{\lambda,\lambda}(x) \quad \text{if} \quad 0 < \mu < \lambda.$$

This gives a characterization of open sets in X_w^* in terms of w:

$$U \in \tau(d_w^0) \quad \Leftrightarrow \quad \forall x \in U \ \exists \mu = \mu(x) > 0 \text{ such that } B_{\mu,\mu}(x) \subset U.$$

$$(4.1.4)$$

The interior of a set $A \subset X_w^*$ is $A^\circ = \{x \in A : B_{\mu,\mu}(x) \subset A \text{ for some } \mu > 0\}$.

Now, suppose w is convex. By virtue of inequalities (2.3.6),

$$B(x, r) \subset B^{d_w^*}(x, r) \subset B(x, \sqrt{r}) \quad \text{for} \quad 0 < r < 1,$$

and so, $\tau(d_w^*) = \tau(d_w^0)$. Noting that $\hat{w}_\lambda(x, y) = \lambda w_\lambda(x, y)$ is a (pseudo)modular on X, $d_w^* = d_{\hat{w}}^0$, and $B_{\lambda,\mu}^{\hat{w}}(x) = B_{\lambda,\mu/\lambda}(x)$ for all $\lambda, \mu > 0$, we get

$$U \in \tau(d_w^*) \quad \Leftrightarrow \quad \forall x \in U \ \exists \lambda = \lambda(x) > 0 \text{ such that } B_{\lambda,1}(x) = B_{\lambda,\lambda}^{\hat{w}}(x) \subset U.$$

3. *Closed sets.* A set $A \subset X_w^*$ is *closed* if its complement $A^c = X_w^* \setminus A$ is open; in other words, $(A^c)^\circ = A^c$ (i.e., A is closed with respect to the metric convergence). Examples of closed sets are $\varnothing$, the *closed ball* of radius $r > 0$ centered at $x \in X_w^*$:

$$\overline{B}(x, r) = \{y \in X_w^* : d_w^0(x, y) \leq r\} = \{y \in X_w^* : w_{r+0}(x, y) \leq r\}$$

(see Theorem 2.2.11(b)), and the modular spaces X_w^*, X_w^0, and X_w^{fin}. The *closure* of a set $A \subset X_w^*$, denoted by $\overline{A}$, is the smallest closed set containing A (i.e., the intersection of all closed sets $F \supset A$). Clearly, $\overline{A} = ((A^c)^\circ)^c$ is closed, $A \subset \overline{A}$, and A is closed if and only if $\overline{A} = A$. Since the closure $\overline{A}$ is the set of points from X_w^*, which do not belong to $(A^c)^\circ$, we have

$$\overline{A} = \{x \in X_w^* : A \cap B_{\mu,\mu}(x) \neq \varnothing \text{ for all } \mu > 0\}.$$

Note that $\overline{B(x, r)} \subset \overline{B}(x, r)$ and $B(x, r) \subset (\overline{B}(x, r))^\circ$ are proper inclusions in general (e.g., $w_\lambda(x, y) = \delta(x, y)$ on X and $B(x, 1) = \{x\}$).

Lemma 4.1.2. *If $\varnothing \neq A \subset X_w^*$ and $x \in X_w^*$, the following are equivalent:*

(a) $x \in \overline{A}$;
(b) *there is a sequence $\{x_n\} \subset A$ such that $w_\lambda(x_n, x) \to 0$ for all $\lambda > 0$;*
(c) *for every $\lambda > 0$ there is a sequence $\{x_n(\lambda)\} \subset A$ such that $w_\lambda(x_n(\lambda), x) \to 0$.*

Proof. (a)$\Leftrightarrow$(b)   This follows from Theorem 4.1.1 if we note that (a) holds if and only if there exists a sequence $\{x_n\} \subset A$ such that $d_w^0(x_n, x) \to 0$.

(b)$\Leftrightarrow$(c)   Clearly, (b) implies (c) with $x_n(\lambda) = x_n$ for all $\lambda > 0$ and $n \in \mathbb{N}$. Now, suppose that (c) holds. Given $k \in \mathbb{N}$, there exists a sequence $\{x_n(1/k)\}_{n=1}^\infty \subset A$ such that $w_{1/k}(x_n(1/k), x) \to 0$ as $n \to \infty$. Choose $n_1 \in \mathbb{N}$ such that $w_1(x_{n_1}(1), x) < 1$. Then, pick $n_2 \in \mathbb{N}$ such that $n_2 > n_1$ and $w_{1/2}(x_{n_2}(1/2), x) < 1/2$. Inductively, if $k \geq 3$ and $n_{k-1} \in \mathbb{N}$ is already chosen, pick $n_k \in \mathbb{N}$ such that $n_k > n_{k-1}$ and $w_{1/k}(x_{n_k}(1/k), x) < 1/k$. Define $y_k = x_{n_k}(1/k)$ for all $k \in \mathbb{N}$, so that $\{y_k\}_{k=1}^\infty \subset A$. We claim that $w_\lambda(y_k, x) \to 0$ as $k \to \infty$ for all $\lambda > 0$. In fact, given $\lambda > 0$, let $k_0 = k_0(\lambda) \in \mathbb{N}$ be such that $k_0 > 1/\lambda$. Then, for all $k \geq k_0$, we find $1/k < \lambda$, and so, by the monotonicity of w, $w_\lambda(y_k, x) \leq w_{1/k}(y_k, x) < 1/k$.
$\square$

We denote by $\mathrm{cl}\,(X_w^*)$ the family of all closed subsets of X_w^*.

Theorem 4.1.3. *If w is a modular on X, then the Hausdorff pseudomodular W on $\mathscr{P}(X)$, induced by w, is a modular on $\mathrm{cl}\,(X_w^*)$, and so (by Theorems 2.2.1 and 2.2.13), $d_W^0 = D_{d_w^0}$ is an extended metric on $\mathrm{cl}\,(X_w^*)$.*

Proof. Taking into account Sect. 1.3.5, it suffices to establish that, given $A, B \subset X_w^*$, we have

$$E_\lambda(A, B) = 0 \text{ for all } \lambda > 0 \quad \text{if and only if} \quad A \subset \overline{B}. \qquad (4.1.5)$$

In fact, it follows from (4.1.5) that if $A, B \in \mathrm{cl}\,(X_w^*)$ and $W_\lambda(A, B) = 0$ for all $\lambda > 0$, then $E_\lambda(A, B) = E_\lambda(B, A) = 0$ implying $A \subset \overline{B} = B$ and $B \subset \overline{A} = A$, i.e., $A = B$.

To prove (4.1.5), we may assume that $A \neq \varnothing$ (which also yields $B \neq \varnothing$).

($\Rightarrow$) If $E_\lambda(A, B) = 0$ for all $\lambda > 0$, then, by (1.3.12), given $x \in A$, we find $\inf_{y \in B} w_\lambda(x, y) = 0$. Hence, for every $\lambda > 0$ there is a sequence $\{y_n(\lambda)\} \subset B$ (also depending on x) such that $w_\lambda(x, y_n(\lambda)) \to 0$ as $n \to \infty$. Lemma 4.1.2 implies $x \in \overline{B}$. Thus, $A \subset \overline{B}$.

($\Leftarrow$) Suppose $A \subset \overline{B}$. Given $x \in A$, we have $x \in \overline{B}$, and so, by Lemma 4.1.2, there is a sequence $\{y_n\} \subset B$ such that $w_\lambda(x, y_n) \to 0$ as $n \to \infty$ for all $\lambda > 0$. Since $\inf_{y \in B} w_\lambda(x, y) \leq w_\lambda(x, y_n) \to 0$ as $n \to \infty$, we get $\inf_{y \in B} w_\lambda(x, y) = 0$ for all $x \in A$ and $\lambda > 0$. By (1.3.12), this gives $E_\lambda(A, B) = 0$ for all $\lambda > 0$. $\qquad\square$

4.2 The Modular Convergence

Let w be a (pseudo)modular on X.

As we have seen in Theorem 4.1.1, the metric convergence in X_w^* is equivalent to the convergence in modular w for *all* parameters $\lambda > 0$. Formally, we get a more general (or weak) convergence if we allow for the right-hand side of (4.1.1) to hold only for *some* $\lambda > 0$. In the sequel, modular entourages $B_{\lambda, \mu}(x)$ play the same significant role for the modular topology as open (closed) balls do for the metric topology (note that the entourages may be neither open, nor closed).

Definition 4.2.1. A sequence $\{x_n\} \subset X$ is said to be *w-convergent* (or *modular convergent*) to $x \in X$ (in symbols, $x_n \overset{w}{\to} x$) if there exists $\lambda_0 > 0$, possibly depending on $\{x_n\}$ and x, such that $w_{\lambda_0}(x_n, x) \to 0$. Any such element x is called a *w-limit* (or *modular limit*) of $\{x_n\}$. (By the monotonicity of w, $w_\lambda(x_n, x) \to 0$ for all $\lambda \geq \lambda_0$.)

Clearly, the metric convergence (w.r.t. d_w^0) implies the modular convergence. Let us illustrate this definition by an example.

Example 4.2.2. Assume that (X, d) is a given metric space.

(1) If $p \geq 0$, let $w_\lambda(x, y) = d(x, y)/\lambda^p$ be the modular from Examples 1.3.2(1) and 2.2.2(1). Since $X_w^* = X$ and $d_w^0(x, y) = \sqrt[p+1]{d(x, y)}$, the metric convergence and modular convergence in X_w^* are equivalent to the usual d-convergence in X.

(2) Now, let w be the modular from (1.3.9) and Example 2.3.5(2): $w_\lambda(x, y) = \infty$ if $0 < \lambda < d(x, y)$, and $w_\lambda(x, y) = 0$ if $\lambda \geq d(x, y)$, so that we have $X_w^* = X_w^0 = X$ and $d_w^0 = d_w^* = d$. A sequence $\{x_n\} \subset X_w^*$, convergent in metric $d_w^0 = d$, is bounded in (X, d) (cf. Remark 2.4.3(1)), but not vice versa. On the other hand, the w-convergence of $\{x_n\} \subset X_w^*$ is equivalent to the boundedness of $\{x_n\}$ in (X, d).

In fact, if $x_n \overset{w}{\to} x \in X$, then $w_{\lambda_0}(x_n, x) \to 0$ for some $\lambda_0 > 0$, and so, there is $n_0 \in \mathbb{N}$ such that, if $n > n_0$, $w_{\lambda_0}(x_n, x) < 1$, which is equivalent to $\lambda_0 \geq d(x_n, x)$. Hence $\{x_n\} \subset \overline{B}(x, r_0)$ with $r_0 = \max\{d(x_1, x), \ldots, d(x_{n_0}, x), \lambda_0\}$. Conversely, if

$\{x_n\} \subset \overline{B}(x,r)$ for some $x \in X$ and $r > 0$, then $x_n \xrightarrow{w} y$ for every $y \in X$: since $d(x_n, y) \leq d(x_n, x) + d(x, y) \leq \lambda_0$ for $n \in \mathbb{N}$ and $\lambda_0 = r + d(x, y)$, we have $w_{\lambda_0}(x_n, y) = 0$ for all $n \in \mathbb{N}$.

Remark 4.2.3. Note that modulars from Example 4.2.2 in (1) for $p = 0$ and (2) are mutually right inverses of each other (see (3.3.3) in Example 3.3.9) and generate the same metric $d_w^0 = d$ on $X_w^* = X$, but their modular convergences are different. See also Example 4.3.4(1) on p. *74.*

This section aims at exhibiting *topological aspects* of the modular convergence.
We say that elements $x, y \in X$ are *w-equal* (or *modular equal*) if there is $\lambda > 0$ such that $w_\lambda(x, y) = 0$ (symbolically: $x \overset{w}{=} y$). By axiom (i_s), given $x, y \in X$,

$$\left(x \overset{w}{=} y \text{ is equivalent to } x = y\right) \quad \text{if and only if} \quad \left(w \text{ is strict}\right).$$

For the modular w from Example 4.2.2(2), we have $x \overset{w}{=} y$ for all $x, y \in X$.
A set $A \subset X$ is said to be *closed with respect to the modular convergence* if, given $\{x_n\} \subset A$ such that $x_n \xrightarrow{w} x \in X$, we have $x \in A$.

Theorem 4.2.4. (a) *Modular spaces X_w^* and X_w^0 (around $x^\circ \in X$) are closed with respect to the modular convergence, while X_w^{fin} is generally not.*
(b) *A modular convergent sequence from X_w^* is bounded (in metric d_w^0), all its modular limits are w-equal, and if w is a strict modular on X, at most one modular limit may exist.*

Proof. (a) Suppose $\{x_n\} \subset X$ and $x_n \xrightarrow{w} x \in X$. Then $w_{\lambda_0}(x_n, x) \to 0$ for some $\lambda_0 > 0$, and so, given $\varepsilon > 0$, there is $n_0 = n_0(\varepsilon) \in \mathbb{N}$ such that $w_{\lambda_0}(x_n, x) \leq \varepsilon$ for all $n \geq n_0$. By axiom (iii), it follows that, if $\lambda_1 > 0$,

$$w_{\lambda_0+\lambda_1}(x, x^\circ) \leq w_{\lambda_0}(x, x_{n_0}) + w_{\lambda_1}(x_{n_0}, x^\circ) \leq \varepsilon + w_{\lambda_1}(x_{n_0}, x^\circ). \qquad (4.2.1)$$

Let $\{x_n\} \subset X_w^*$. Since $x_{n_0} \in X_w^*$, $w_{\lambda_1}(x_{n_0}, x^\circ) < \infty$ for some $\lambda_1 > 0$, and so, (4.2.1) implies $w_{\lambda_0+\lambda_1}(x, x^\circ) < \infty$, i.e., $x \in X_w^*$.
If $\{x_n\} \subset X_w^0$, then $x_{n_0} \in X_w^0$, and so, $w_\lambda(x_{n_0}, x^\circ) \to 0$ as $\lambda \to \infty$. Choose $\lambda_1 = \lambda_1(\varepsilon) > 0$ such that $w_{\lambda_1}(x_{n_0}, x^\circ) \leq \varepsilon$. Hence, if $\lambda \geq \lambda_0 + \lambda_1$, (4.2.1) implies $w_\lambda(x, x^\circ) \leq w_{\lambda_0+\lambda_1}(x, x^\circ) \leq 2\varepsilon$, i.e., $w_\lambda(x, x^\circ) \to 0$ as $\lambda \to \infty$, and $x \in X_w^0$.
The corresponding assertion for X_w^{fin} is considered in Example 4.2.7(2).

(b) Suppose $\{x_n\} \subset X_w^*$ and $x_n \xrightarrow{w} x \in X$. By (a), $x \in X_w^*$, and there is $n_0 \in \mathbb{N}$ such that $w_{\lambda_0}(x_n, x) \leq \lambda_0$ for all $n > n_0$, which implies $d_w^0(x_n, x) \leq \lambda_0$. By Theorem 2.2.1, d_w^0 assumes finite values on pairs of elements from X_w^*, and so, $\{x_n\} \subset \overline{B}(x, r)$, where $r = \max\{d_w^0(x_1, x), \ldots, d_w^0(x_{n_0}, x), \lambda_0\}$.
If $x_n \xrightarrow{w} x$ and $x_n \xrightarrow{w} y$, then $w_\lambda(x_n, x) \to 0$ and $w_\mu(x_n, y) \to 0$ for some $\lambda, \mu > 0$. By axiom (iii), $w_{\lambda+\mu}(x, y) \leq w_\lambda(x, x_n) + w_\mu(x_n, y) \to 0$, and so, $w_{\lambda+\mu}(x, y) = 0$. Thus, $x \overset{w}{=} y$, and if, in addition, w is strict, then $x = y$. $\qquad \square$

Definition 4.2.5. We say that w *satisfies the Δ_2-condition* on X, or simply w *is Δ_2* on X, if, given a sequence $\{x_n\} \subset X$, $x \in X$ and $\lambda > 0$ such that $w_\lambda(x_n, x) \to 0$, we have $w_{\lambda/2}(x_n, x) \to 0$. A similar definition applies with X replaced by X_w^*.

Theorem 4.2.6. (a) *The w-convergence on X is equivalent to the metric convergence if and only if w is Δ_2 on X.*

(b) *The (pseudo)modular w is not Δ_2 on X if and only if there are a sequence $\{x_n\} \subset X$, $x \in X$, and $\varepsilon_0 > 0$ such that $x_n \overset{w}{\to} x$, and $d_w^0(x_n, x) \geq \varepsilon_0$ for all $n \in \mathbb{N}$.*

(c) *If w is a nonstrict modular on X, then w is not Δ_2 on X.*

Proof. (a)($\Rightarrow$) Let $\{x_n\} \subset X$, $x \in X$, and $\lambda_0 > 0$ be such that $w_{\lambda_0}(x_n, x) \to 0$. We have $x_n \overset{w}{\to} x$, and so, by the assumption, $d_w^0(x_n, x) \to 0$. Theorem 4.1.1(a) implies in particular $w_{\lambda_0/2}(x_n, x) \to 0$, i.e., w is Δ_2 on X.

(a)($\Leftarrow$) We need to show only that the w-convergence implies the metric one. Assume $\{x_n\} \subset X$ and $x_n \overset{w}{\to} x \in X$. Then $w_{\lambda_0}(x_n, x) \to 0$ for some $\lambda_0 > 0$, and so, by the Δ_2-condition, $w_{\lambda_0/2}(x_n, x) \to 0$. Inductively, $w_{\lambda_0/2^k}(x_n, x) \to 0$ as $n \to \infty$ for all $k \in \mathbb{N}$. Given $\lambda > 0$, let $k \in \mathbb{N}$ be such that $\lambda > \lambda_0/2^k$. The monotonicity of w implies $w_\lambda(x_n, x) \leq w_{\lambda_0/2^k}(x_n, x) \to 0$ as $n \to \infty$. Thus, $w_\lambda(x_n, x) \to 0$ for all $\lambda > 0$, and so, by Theorem 4.1.1(a), $x_n \to x$ (i.e., $d_w^0(x_n, x) \to 0$).

(b) If w is not Δ_2, then the w-convergence is not equivalent to the d_w^0-convergence, and so, there exist a sequence $\{y_k\} \subset X$ and $x \in X$ such that $y_k \overset{w}{\to} x$ and $y_k \not\to x$. So, for some $\lambda_0 > 0$ and $\varepsilon_0 > 0$, $w_{\lambda_0}(y_k, x) \to 0$, and there is an increasing sequence $\{k_n\}_{n=1}^\infty \subset \mathbb{N}$ such that $d_w^0(y_{k_n}, x) \geq \varepsilon_0$ for all $n \in \mathbb{N}$. The sequence $\{x_n\} = \{y_{k_n}\}_{n=1}^\infty$ has the desired properties.

Conversely, the assumption on the right implies that the w-convergence and metric convergence are not equivalent, and so, by (a), w is not Δ_2 on X.

(c) By the negation of condition (i_s), there are $x, y \in X$, $x \neq y$, and $\lambda_0 > 0$ such that $w_{\lambda_0}(x, y) = 0$. If $\mu = \inf\{\lambda > 0 : w_\lambda(x, y) = 0\}$, axiom (i) implies $\mu > 0$ (otherwise, if $\mu = 0$, then we have $x = y$). Letting $x_n = x$ for all $n \in \mathbb{N}$, and $\lambda = 3\mu/2$, we get $w_\lambda(x_n, y) = w_{3\mu/2}(x, y) = 0$ for all $n \in \mathbb{N}$, and so, $x_n \overset{w}{\to} y$; at the same time, $w_{\lambda/2}(x_n, y) = w_{3\mu/4}(x, y) \neq 0$ for all $n \in \mathbb{N}$. $\qquad\square$

Example 4.2.7. (1) Let $\varphi : [0, \infty) \to [0, \infty)$ be a nondecreasing function such that $\varphi(+0) = 0$, $\varphi(u) > 0$ for $u > 0$, and $\varphi(\infty) = \infty$. Then, φ *satisfies the Δ_2-condition* if there is a constant $c > 0$ such that $\varphi(2u) \leq c\varphi(u)$ for all $u \geq 0$. For instance, functions $\varphi(u) = u^p$ $(p > 0)$ and $\varphi(u) = (1 + |\log u|)u^p$ (for $p \geq 1$) satisfy the Δ_2-condition, while functions $\varphi(u) = \exp(u^p) - 1$ with $p \geq 1$ do not.

If φ satisfies the Δ_2-condition, then the modular w from (1.3.10) with $h(\lambda) = \lambda$ is Δ_2 on $X = M^\mathbb{N}$, which is a consequence of the inequality $w_{\lambda/2}(x, y) \leq cw_\lambda(x, y)$ for all $\lambda > 0$ and $x, y \in X$.

(2) Modular w from (2.4.1) is Δ_2 on $X = M^\mathbb{N}$; in fact, by virtue of (2.4.2), we have $w_{\lambda/2}(x, y) \leq 2w_\lambda(x, y)$. We are going to show that its right inverse w^+, which is given by (3.3.9), is not Δ_2 on X. Below we present a sequence $\{x^{(k)}\}_{k=1}^\infty \subset X_{w^+}^{\text{fin}}$

with the properties: $x^{(k)} \xrightarrow{w^+} x \notin X^{\text{fin}}_{w^+}$ as $k \to \infty$ (which means that the modular space $X^{\text{fin}}_{w^+}$ is *not closed* with respect to the w^+-convergence), and $d^0_{w^+}(x^{(k)}, x) \not\to 0$ as $k \to \infty$ (by Theorem 4.2.6(a), this implies that w^+ is not Δ_2).

We set $M = \mathbb{R}$, and so, w^+ is of the form

$$w^+_\mu(x, y) = \sup_{n \in \mathbb{N}} \frac{|x_n - y_n|}{\mu^n}, \quad \mu > 0, \quad x = \{x_n\}, \ y = \{y_n\} \in X = \mathbb{R}^{\mathbb{N}}.$$

For the center $x^\circ = 0 = \{0\}^\infty_{n=1}$, embeddings (3.3.11) and Proposition 2.4.2 imply

$$X^{\text{fin}}_{w^+} = X^0_w = \left\{ x = \{x_n\} \in X : \lim_{n \to \infty} |x_n|^{1/n} = 0 \right\} \subset X^*_{w^+} = X^0_{w^+}.$$

If $k \in \mathbb{N}$, we set $x^{(k)} = (1/k, 1_2, 1_3, \ldots, 1_k, 0, 0, \ldots)$ (in other words, $x^{(k)} = \{x_n^{(k)}\}^\infty_{n=1}$ with $x_1^{(k)} = 1/k$, $x_n^{(k)} = 1$ for $2 \le n \le k$, and $x_n^{(k)} = 0$ for all $n \ge k+1$). Noting that a w^+-convergent sequence converges also coordinatewise, we define the limit x by $x = \{x_n\}^\infty_{n=1} = (0, 1, 1, 1, \ldots)$. Clearly, all $x^{(k)}$ belong to $X^{\text{fin}}_{w^+}$, whereas, by Theorem 3.3.8(a) and (2.4.3), $x \in X^*_{w^+} = X^*_w$, and $x \notin X^{\text{fin}}_{w^+}$. If $\mu = 2$, we get

$$\left\{ \frac{|x_n^{(k)} - x_n|}{\mu^n} \right\}^\infty_{n=1} = \left(\frac{1}{2k}, 0_2, \ldots, 0_k, \frac{1}{2^{k+1}}, \frac{1}{2^{k+2}}, \frac{1}{2^{k+3}}, \ldots \right)$$

and, since $1/2k > 1/2^{k+1} > 1/2^{k+2} > \ldots$, we find $w^+_2(x^{(k)}, x) = 1/2k$, i.e., $x^{(k)} \xrightarrow{w^+} x$.

On the other hand, by Theorem 3.3.8(c), metric $d^0_{w^+} = d^0_w$ is given by (2.4.6), whence $d^0_{w^+}(x^{(k)}, x) = \sup_{n \in \mathbb{N}} |x_n^{(k)} - x_n|^{1/(n+1)} = 1$ for all $k \in \mathbb{N}$. (The assertion $d^0_{w^+}(x^{(k)}, x) \not\to 0$ also follows indirectly from Theorem 4.1.1(c).)

Remark 4.2.8. The sequence $\{x^{(k)}\}$ from the Example above does not w-converge to x with respect to the original modular w from (2.4.1): if this was not so, Theorem 4.2.4(a) would imply $x \in X^0_w = X^{\text{fin}}_{w^+}$, which contradicts $x \notin X^{\text{fin}}_{w^+}$. Furthermore, the sequence $\{x^{(k)}\}$ does not converge to x uniformly: since $\{|x_n^{(k)} - x_n|\}^\infty_{n=1} = (1/k, 0_2, \ldots, 0_k, 1, 1, 1, \ldots)$, we find $\sup_{n \in \mathbb{N}} |x_n^{(k)} - x_n| = 1$ for all $k \in \mathbb{N}$. Thus, the w-convergence in $X^*_{w^+}$ is intermediate between the uniform and coordinatewise ones: (uniform convergence) $\Rightarrow$ (w-convergence) $\Rightarrow$ (coordinatewise convergence).

4.3 The Modular Topology

We begin with the definition of modular open sets and their properties.

Definition 4.3.1. A nonempty set $U \subset X$ is said to be w-*open* (or *modular open*) if for every $x \in U$ and $\lambda > 0$ there is $\mu > 0$ (possibly depending on x and λ) such that $B_{\lambda,\mu}(x) \subset U$.

Note that if $x \in X_w^*(x^\circ)$, then $B_{\lambda,\mu}(x) \subset X_w^*(x^\circ)$ for all $\lambda, \mu > 0$. It follows that $U \subset X$ is w-open if and only if $U \cap X_w^*(x^\circ)$ is w-open for all $x^\circ \in X$. So, in the sequel we consider w-open subsets of X_w^* (with a fixed center $x^\circ \in X$).

Denote by $\tau(w)$ the family of all w-open subsets of X_w^*. Clearly, $\tau(w)$ is a topology on X_w^* (i.e., $\tau(w)$ contains $\varnothing$ and X_w^* and is closed under arbitrary unions and finite intersections). It is called the w-*topology* (or *modular topology*) on X_w^*.

Nontrivial examples of w-open sets are given by the following Lemma.

Lemma 4.3.2. *Let* $U_\varphi(x) = \bigcup_{\lambda > 0} B_{\lambda,\varphi(\lambda)}(x)$, *where* $x \in X_w^*$ *and* $\varphi : (0, \infty) \to (0, \infty)$ *is a function. Then* $U_\varphi(x) \in \tau(w)$, *provided one of the following two conditions hold:*

(a) φ *is nondecreasing on* $(0, \infty)$;
(b) w *is convex and* $\lambda \mapsto \lambda\varphi(\lambda)$ *is nondecreasing on* $(0, \infty)$.

Proof. In both cases we have to verify that, given $y \in U_\varphi(x)$ and $\lambda > 0$, there is a number $\mu = \mu(y, \lambda) > 0$ such that $B_{\lambda,\mu}(y) \subset U_\varphi(x)$. Since $y \in U_\varphi(x)$, there exists $\lambda_0 = \lambda_0(y) > 0$ such that $y \in B_{\lambda_0,\varphi(\lambda_0)}(x)$, i.e., $w_{\lambda_0}(x, y) < \varphi(\lambda_0)$.

(a) Set $\mu = \mu(y, \lambda) = \varphi(\lambda_0 + \lambda) - w_{\lambda_0}(x, y)$. By the monotonicity of φ, μ is well-defined, i.e., $\mu \geq \varphi(\lambda_0) - w_{\lambda_0}(x, y) > 0$. Now, let $z \in B_{\lambda,\mu}(y)$. Then $w_\lambda(y, z) < \mu$ and, by axiom (iii), we find

$$w_{\lambda_0 + \lambda}(x, z) \leq w_{\lambda_0}(x, y) + w_\lambda(y, z) < w_{\lambda_0}(x, y) + \mu = \varphi(\lambda_0 + \lambda),$$

and so, $z \in B_{\lambda_0 + \lambda, \varphi(\lambda_0 + \lambda)}(x) \subset U_\varphi(x)$. This shows that $B_{\lambda,\mu}(y) \subset U_\varphi(x)$.

(b) Since the function $\lambda \mapsto \lambda\varphi(\lambda)$ is nondecreasing on $(0, \infty)$, we have

$$(\lambda_0 + \lambda)\varphi(\lambda_0 + \lambda) - \lambda_0 w_{\lambda_0}(x, y) \geq \lambda_0\varphi(\lambda_0) - \lambda_0 w_{\lambda_0}(x, y) > 0,$$

and so, the following quantity is positive:

$$\mu = \mu(y, \lambda) = \frac{(\lambda_0 + \lambda)\varphi(\lambda_0 + \lambda) - \lambda_0 w_{\lambda_0}(x, y)}{\lambda}.$$

If $z \in B_{\lambda,\mu}(y)$, then $w_\lambda(y, z) < \mu$, whence, by axiom of convexity (iv),

$$w_{\lambda_0+\lambda}(x, z) \leq \frac{\lambda_0}{\lambda_0+\lambda} w_{\lambda_0}(x, y) + \frac{\lambda}{\lambda_0+\lambda} w_\lambda(y, z) < \frac{\lambda_0 w_{\lambda_0}(x, y) + \lambda \mu}{\lambda_0+\lambda} = \varphi(\lambda_0+\lambda).$$

Thus, $z \in B_{\lambda_0+\lambda,\varphi(\lambda_0+\lambda)}(x) \subset U_\varphi(x)$, which proves that $B_{\lambda,\mu}(y) \subset U_\varphi(x)$. □

Remark 4.3.3. It follows from Lemma 4.3.2 that $U_{x,\varepsilon} = \bigcup_{\lambda>0} B_{\lambda,\varepsilon}(x) \in \tau(w)$ for all $\varepsilon > 0$, and if w is convex, then $U^*_{x,\varepsilon} = \bigcup_{\lambda>0} B_{\lambda,\varepsilon/\lambda}(x) \in \tau(w)$ for all $\varepsilon > 0$. Thus, modular open sets are rather 'massive' (cf. (4.1.4)). Note that the family $\{U_{x,\varepsilon} : \varepsilon > 0\}$ may not form a neighborhood base for $\tau(w)$ at the point $x \in X^*_w$ (see Example 4.3.4(1) concerning modular $w_\lambda(x, y) = d(x, y)/\lambda^p$).

Example 4.3.4. (1) We denote by $B^d(x, r)$ and $\overline{B}^d(x, r)$ the usual open and closed balls in a metric space (X, d), respectively.

In Example 4.2.2(1), we have, for $r > 0$ and $\lambda, \mu > 0$,

$$B(x, r) = \{y \in X : d^0_w(x, y) < r\} = B^d(x, r^{p+1})$$

and

$$B_{\lambda,\mu}(x) = \{y \in X : w_\lambda(x, y) < \mu\} = B^d(x, \lambda^p \mu).$$

Since $B_{\lambda,\mu}(x) = B(x, r)$ with $\mu = r^{p+1}/\lambda^p$, we find $\tau(w) = \tau(d^0_w)$.

In Example 4.2.2(2), we have $B(x, r) = B^d(x, r)$ and $B_{\lambda,\mu}(x) = \overline{B}^d(x, \lambda)$. Therefore, $\varnothing \neq U \in \tau(w)$ iff $\overline{B}^d(x, \lambda) \subset U$ for all $x \in U$ and $\lambda > 0$, which is equivalent to $U = X$. Hence, the w-topology $\tau(w) = \{\varnothing, X\}$ is the *antidiscrete* (nonmetrizable!) topology on $X^*_w = X$, which is a proper subfamily of the metric topology $\tau(d^0_w) = \tau(d)$. Note that the w-convergence $x_n \overset{w}{\to} x$ implies the convergence $x_n \overset{\tau(w)}{\to} x$ in topology $\tau(w)$, but not vice versa: in fact, by Example 4.2.2(2), only bounded sequences in (X, d) are w-convergent, whereas all sequences in (X, d) are convergent in the antidiscrete topology $\tau(w)$.

(2) Suppose w is the modular from Example 2.3.5(5). We have $X^*_w = X$, and the modular entourages for w are given by

$$B^w_{\lambda,\mu}(x) = \begin{cases} \{x\} & \text{if } \lambda > 0 \text{ and } 0 < \mu \leq 1, \\ B^d(x, \lambda\mu) & \text{if } \lambda > 0 \text{ and } \quad \mu > 1. \end{cases}$$

Since $d^0_w(x, y) = \max\{1, \sqrt{d(x, y)}\}$ if $x \neq y$ (and $= 0$ if $x = y$), the metric balls in (X^*_w, d^0_w) are of the form

$$B(x, r) = \{y \in X : d^0_w(x, y) < r\} = \begin{cases} \{x\} & \text{if } 0 < r \leq 1, \\ B^d(x, r^2) & \text{if } \quad r > 1. \end{cases}$$

Hence $\tau(w) = \tau(d^0_w) = \mathscr{P}(X)$ is the *discrete* topology on X.

The right inverse w^+ of w is given by (3.3.7), $X^*_{w^+} = X$ and, since $d^0_{w^+} = d^0_w$, the metric balls with respect to $d^0_{w^+}$ are the same as above. Note that the modular entourages for w^+ are as follows:

$$B^{w^+}_{\lambda,\mu}(x) = \begin{cases} \{x\} & \text{if } 0 < \lambda < 1 \text{ and } \mu > 0, \\ B^d(x, \lambda\mu) & \text{if } \lambda \geq 1 \text{ and } \mu > 0. \end{cases}$$

So, $\tau(w^+) = \tau(d)$ is the metric topology on (X, d). It is to be noted that the topology $\tau(w^+)$ is metrizable, but not by means of metric $d^0_{w^+}$.

Theorem 4.3.5. (a) $\tau(w) \subset \tau(d^0_w)$ (i.e., every w-open set is open);
(b) $\tau(w) = \tau(d^0_w)$ iff w is Δ_2 on X^*_w;
(c) $\tau(w)$ is the finest topology among all topologies τ on X^*_w such that, given $\{x_n\} \subset X^*_w$ and $x \in X^*_w$, $x_n \xrightarrow{w} x$ implies $x_n \xrightarrow{\tau} x$.

Proof. (a) Let $U \in \tau(w)$. For any $x \in U$ and $\lambda = 1$, there is $\mu_0 = \mu_0(x, 1) > 0$ such that $B_{1,\mu_0}(x) \subset U$. If $\mu = \min\{1, \mu_0\}$, then $B_{\mu,\mu}(x) \subset B_{1,\mu_0}(x) \subset U$, and so, by virtue of (4.1.4), we get $U \in \tau(d^0_w)$.

(c) Now, we prove (c). First, we show that $x_n \xrightarrow{w} x$ implies $x_n \xrightarrow{\tau(w)} x$. Let $U \in \tau(w)$ be such that $x \in U$. Since $x_n \xrightarrow{w} x$, $w_{\lambda_0}(x_n, x) \to 0$ for some $\lambda_0 > 0$, and since U is w-open, there exists $\mu_0 = \mu_0(x, \lambda_0) > 0$ such that $B_{\lambda_0,\mu_0}(x) \subset U$. Pick $n_0 = n_0(\mu_0) \in \mathbb{N}$ such that $w_{\lambda_0}(x_n, x) < \mu_0$ for all $n \geq n_0$. It follows that $\{x_n : n \geq n_0\} \subset B_{\lambda_0,\mu_0}(x) \subset U$, and so, $x_n \xrightarrow{\tau(w)} x$.

Second, we show that if τ is a topology on X^*_w such that $x_n \xrightarrow{w} x$ implies $x_n \xrightarrow{\tau} x$, then $\tau \subset \tau(w)$ (i.e., $\tau(w)$ is the *finest* among the topologies τ). On the contrary, assume that $U \notin \tau(w)$ for some $U \in \tau$. Then, there are $x_0 \in U$ and $\lambda_0 > 0$ such that for every $n \in \mathbb{N}$ there exists $x_n \in B_{\lambda_0,1/n}(x_0)$ with $x_n \notin U$. From inequality $w_{\lambda_0}(x_n, x_0) < 1/n$, we get $x_n \xrightarrow{w} x_0$, and so, by the assumption, $x_n \xrightarrow{\tau} x_0$. Since $U \in \tau$ and $x_0 \in U$, there exists $n_0 \in \mathbb{N}$ such that $x_n \in U$ for all $n \geq n_0$, which contradicts the condition $\{x_n : n \in \mathbb{N}\} \subset X^*_w \setminus U$.

(b)($\Rightarrow$) By Theorem 4.2.6(a), it suffices to show that the modular convergence implies the metric convergence. Let $\{x_n\} \subset X^*_w$ and $x_n \xrightarrow{w} x \in X^*_w$. Item (c) above implies $x_n \xrightarrow{\tau(w)} x$ and, since $\tau(w) = \tau(d^0_w)$ is the metric topology on X^*_w, $d^0_w(x_n, x) \to 0$.

(b)($\Leftarrow$) By virtue of item (a), we show only that $\tau(d^0_w) \subset \tau(w)$. On the contrary, assume that there is $U \in \tau(d^0_w)$ such that $U \notin \tau(w)$. Then, there are $x_0 \in U$ and $\lambda_0 > 0$ such that, if $n \in \mathbb{N}$, there is $x_n \in B_{\lambda_0,1/n}(x_0)$ such that $x_n \notin U$. It follows that $\{x_n\} \subset X^*_w \setminus U$ and $w_{\lambda_0}(x_n, x_0) < 1/n$ for all $n \in \mathbb{N}$. Hence $x_n \xrightarrow{w} x_0$ and, since w is Δ_2 on X^*_w, Theorem 4.2.6(a) implies $x_n \to x_0$ (in metric d^0_w). Noting that $X^*_w \setminus U$ is closed, we find $x_0 \in X^*_w \setminus U$, which contradicts $x_0 \in U$. $\square$

Definition 4.3.6. The *w-interior* (or *modular interior*) of a set $A \subset X_w^*$, denoted by $A^{\circ w}$, is the largest *w*-open subset of X_w^* contained in A; in other words,

$$A^{\circ w} = \{x \in A : x \in U \subset A \text{ for some } U \in \tau(w)\}.$$

Clearly, $\varnothing^{\circ w} = \varnothing$, $(X_w^*)^{\circ w} = X_w^*$, $A^{\circ w} \in \tau(w)$ and $A^{\circ w} \subset A^{\circ} \subset A$ for all $A \subset X_w^*$ (see Theorem 4.3.5(a) and Sect. 4.1.2), and $A \in \tau(w)$ iff $A^{\circ w} = A$. The operation of taking the modular interior has the usual properties: given $A, B \subset X_w^*$, we have

$$A \subset B \Rightarrow A^{\circ w} \subset B^{\circ w}, \quad (A^{\circ w})^{\circ w} = A^{\circ w}, \quad (A \cap B)^{\circ w} = A^{\circ w} \cap B^{\circ w}, \quad A^{\circ w} \cup B^{\circ w} \subset (A \cup B)^{\circ w}.$$

Definition 4.3.7. A set $A \subset X_w^*$ is said to be *w-closed* (or *modular closed*) if its complement $A^c = X_w^* \setminus A$ is *w*-open (i.e., $A^c \in \tau(w)$), or equivalently, if $(A^c)^{\circ w} = A^c$.

Theorem 4.3.8. (a) *A set $A \subset X_w^*$ is modular closed if and only if A is closed with respect to the modular convergence.*

(b) *Every w-closed subset of X_w^* is closed (but not vice versa), and the converse is true if and only if w is Δ_2 on X_w^*.*

Proof. (a)($\Rightarrow$) Let A be *w*-closed. We must show that, given $\{x_n\} \subset A$ and $x \in X_w^*$ such that $x_n \overset{w}{\to} x$, we have $x \in A$. On the contrary, assume that $x \notin A$, i.e., $x \in A^c = X_w^* \setminus A$. Since $x_n \overset{w}{\to} x$, we get $w_{\lambda_0}(x_n, x) \to 0$ for some $\lambda_0 > 0$. From the *w*-openness of A^c and $x \in A^c$, there is $\mu_0 > 0$ such that $B_{\lambda_0, \mu_0}(x) \subset A^c$. Let $n_0 \in \mathbb{N}$ be such that $w_{\lambda_0}(x_n, x) < \mu_0$ for all $n \geq n_0$. Then $\{x_n : n \geq n_0\} \subset B_{\lambda_0, \mu_0}(x) \subset A^c$, which contradicts the assumption $\{x_n\} \subset A$.

(a)($\Leftarrow$) Now, let A be closed with respect to the *w*-convergence. To show that $A^c \in \tau(w)$, we assume the contrary. Then, there are $x_0 \in A^c$ and $\lambda_0 > 0$ such that, if $n \in \mathbb{N}$, then we can find $x_n \in B_{\lambda_0, 1/n}(x_0)$ such that $x_n \notin A^c$. Hence $\{x_n\} \subset A$ and $x_n \overset{w}{\to} x_0$. It follows that $x_0 \in A$, which contradicts the condition $x_0 \in A^c$.

(b) If $A \subset X_w^*$ is *w*-closed, then its complement A^c is *w*-open. By Theorem 4.3.5(a), A^c is open, and so (Sect. 4.1.2.3), A is closed.

Theorems 4.1.1(c), 4.2.4(a), and 4.3.8(a) imply that the modular spaces X_w^*, X_w^0, and X_w^{fin} are closed, X_w^* and X_w^0 are *w*-closed, while X_w^{fin} is not *w*-closed in general. One more example of a closed set, which is not *w*-closed, is given in step ($\Rightarrow$) below.

($\Rightarrow$) Suppose each closed subset of X_w^* is *w*-closed, but w is not Δ_2 on X_w^*. By Theorem 4.2.6(b), there are a sequence $\{x_n\} \subset X_w^*$, $x \in X_w^*$, and $\varepsilon_0 > 0$ such that $x_n \overset{w}{\to} x$, and $d_w^0(x_n, x) \geq \varepsilon_0$ for all $n \in \mathbb{N}$. If $A = \{x_n : n \in \mathbb{N}\}$ and $\overline{A}$ is the (metric) closure of A, then $\overline{A}$ is closed. However, $\overline{A}$ is not *w*-closed: in fact, $\{x_n\} \subset A \subset \overline{A}$ and $x_n \overset{w}{\to} x$, whereas $x \notin \overline{A}$. This contradicts the assumption.

($\Leftarrow$) Suppose w is Δ_2 on X_w^*. Let $A \subset X_w^*$ be closed, i.e., $A^c \in \tau(d_w^0)$. By Theorem 4.3.5(b), $\tau(d_w^0) = \tau(w)$, and so, $A^c \in \tau(w)$, which means that A is *w*-closed. $\qquad\square$

Definition 4.3.9. The *w-closure* (or *modular closure*) of a set $A \subset X_w^*$, denoted by $\overline{A}^w = A^{-w}$, is the smallest w-closed subset of X_w^* containing A. Clearly,

$$\overline{A}^w = \{x \in X_w^* : A \cap U \neq \varnothing \text{ for all } U \in \tau(w) \text{ with } x \in U\},$$

or, equivalently, $A^{-w} = ((A^c)^{\circ w})^c$, where $A^c = X_w^* \setminus A$.

We have $\overline{\varnothing}^w = \varnothing$, $(X_w^*)^{-w} = X_w^*$, $\overline{A}^w$ is w-closed and $A \subset \overline{A} \subset \overline{A}^w$ for all $A \subset X_w^*$, and A is w-closed iff $\overline{A}^w = A$. The operation of taking the modular closure has the usual properties: given $A, B \subset X_w^*$, we have

$$A \subset B \Rightarrow \overline{A}^w \subset \overline{B}^w, \quad (A^{-w})^{-w} = A^{-w}, \quad \overline{(A \cup B)}^w = \overline{A}^w \cup \overline{B}^w, \quad \overline{(A \cap B)}^w \subset \overline{A}^w \cap \overline{B}^w.$$

Given $A \subset X_w^*$, let $\ell^w(A)$ designate the set of all modular limits of all modular convergent sequences from A:

$$\ell^w(A) = \{x \in X_w^* : x_n \xrightarrow{w} x \text{ for some sequence } \{x_n\} \subset A\}.$$

Theorem 4.3.10. (a) $\ell^w(A)$ *is the set of all elements* $x \in X_w^*$, *for which there exists* $\lambda = \lambda(x) > 0$, *such that* $A \cap B_{\lambda,\mu}(x) \neq \varnothing$ *for all* $\mu > 0$;
(b) $A \subset \overline{A} \subset \ell^w(A) \subset \overline{A}^w$;
(c) $\overline{A} = \overline{A}^w$ *for all* $A \subset X_w^*$ *if and only if* w *is* Δ_2 *on* X_w^*;
(d) *if* $x \in \overline{A}^w$, *then there are sequences* $\{\lambda_n\} \subset (0, \infty)$ *and* $\{x_n\} \subset A$ *such that* $w_{\lambda_n}(x_n, x) \to 0$ *as* $n \to \infty$.

Proof. (a) Suppose $x \in \ell^w(A)$. There are a sequence $\{x_n\} \subset A$ and $\lambda > 0$ such that $w_\lambda(x_n, x) \to 0$. So, given $\mu > 0$, there exists $n_0 = n_0(\mu) \in \mathbb{N}$ such that $w_\lambda(x_n, x) < \mu$ for all $n \geq n_0$, i.e., $\{x_n : n \geq n_0\} \subset B_{\lambda,\mu}(x)$. Hence $A \cap B_{\lambda,\mu}(x) \neq \varnothing$ for all $\mu > 0$.

Conversely, if $x \in X_w^*$ and, for some $\lambda > 0$, $A \cap B_{\lambda,\mu}(x) \neq \varnothing$ for all $\mu > 0$, then for every $n \in \mathbb{N}$ choose an element $x_n \in A \cap B_{\lambda,1/n}(x)$. It follows that $\{x_n\} \subset A$, and $w_\lambda(x_n, x) < 1/n$. Thus, $x \in \ell^w(A)$.

(b) The second inclusion is a consequence of Lemma 4.1.2(b) and the definition of the modular convergence. To obtain the third inclusion, let $x \notin \overline{A}^w$. Since $x \in (\overline{A}^w)^c$ and $(\overline{A}^w)^c$ is w-open, for every $\lambda > 0$ there is $\mu > 0$ such that $B_{\lambda,\mu}(x) \subset (\overline{A}^w)^c \subset A^c$, and so, $A \cap B_{\lambda,\mu}(x) = \varnothing$. By item (a), $x \notin \ell^w(A)$.

(c)($\Rightarrow$) If w is not Δ_2 on X_w^*, then, for the set A from step ($\Rightarrow$) of the proof of Theorem 4.3.8(b), we find $x \in \ell^w(A) \subset \overline{A}^w$ and $x \notin \overline{A}$, and so, $\overline{A} \neq \overline{A}^w$.

(c)($\Leftarrow$) If w is Δ_2 on X_w^*, then, by Theorem 4.3.5(b), $\tau(w) = \tau(d_w^0)$, and so, $(A^c)^\circ = (A^c)^{\circ w}$, which implies $\overline{A} = \overline{A}^w$.

(d) Since $x \in \overline{A}^w$, $A \cap U \neq \varnothing$ for all $U \in \tau(w)$ with $x \in U$. Taking into account Remark 4.3.3, given $n \in \mathbb{N}$, the set $U_{x,1/n} = \bigcup_{\lambda > 0} B_{\lambda,1/n}(x)$ is w-open and $x \in U_{x,1/n}$. Pick $x_n \in A \cap U_{x,1/n}$, i.e., $x_n \in A$ and $x_n \in B_{\lambda_n,1/n}(x)$ for some $\lambda_n > 0$. It follows that $\{x_n\} \subset A$ and $w_{\lambda_n}(x_n, x) < 1/n$ for all $n \in \mathbb{N}$. $\qquad\square$

Remark 4.3.11. (1) In general, $\ell^w(A)$ is not w-closed, and so, $\ell^w(A) \neq \overline{A}^w$.

(2) If the modular topology $\tau(w)$ is Hausdorff, then w is strict. In fact, suppose $x, y \in X_w^*$ are such that $x \neq y$, and $\lambda > 0$. There are $U_x, U_y \in \tau(w)$ such that $x \in U_x$, $y \in U_y$, and $U_x \cap U_y = \emptyset$. By the w-openness of U_x, there exists $\mu = \mu(x, \lambda) > 0$ such that $B_{\lambda,\mu}(x) \subset U_x$. Hence $y \notin B_{\lambda,\mu}(x)$, i.e., $w_\lambda(x, y) \geq \mu > 0$. By the definition (i$_s$), w is a strict modular.

4.4   Bibliographical Notes and Comments

Section 4.1.   Theorem 4.1.1 was established in Chistyakov [24, Theorem 2.13] and [28, Theorems 1 and 2]. It is a natural extension of the corresponding result for modulars on linear spaces in Musielak [75, Theorem 1.6]. In Sect. 4.1.2, we apply the usual metric space terminology to the (pseudo)metric modular space (X_w^*, d_w^0) (and (X_w^*, d_w^*) for convex w) and characterize metrically open and closed sets in X_w^* in terms of w. Modular entourages are good candidates for the description of topologies on X_w^*. Lemma 4.1.2(c) and Theorem 4.1.3 are new.

Section 4.2.   The notion of modular convergence for classical modulars was introduced by Musielak and Orlicz [77] (see Musielak [75, Chap. I, Sect. 5]). Since our modulars are extensions of classical modulars, we apply the same terminology as in the classical modular spaces theory (closed with respect to the modular convergence, Δ_2-condition) with suitable modifications. For metric modulars w, these notions were considered in Chistyakov [28], where Theorems 4.2.4 and 4.2.6 were established. Example 4.2.7(2) is new.

Section 4.3.   The modular topology on linear modular spaces was investigated by Leśniewicz [60, 61] and Leśniewicz and Orlicz [63] (see also Musielak [75, Chap. I, Sect. 6], and Nowak [84, 85]). Since modulars in our sense convey an impression of 'distorted' metrics (to a certain extent), we have attempted an approach to the modular topology on X_w^* similar to the one adopted for metric spaces: a modular closed set (i.e., the complement to a modular open set) must be closed with respect to the modular convergence, and vice versa. This led exactly to Definition 4.3.1, and Theorem 4.3.8. Although the material of Sect. 4.3 is new in the context of metric modulars w, it is quite standard from the point of view of the general topology.

Chapter 5
Bounded and Regulated Mappings

Abstract In this chapter, we introduce and study a special $\mathbb{N}$-valued modular on the set of all mappings from an interval of the real line into a metric space. We show that the sets of all *bounded* mappings and *regulated* mappings (i.e., those, whose one-sided limits exist at each point of the interval) are modular spaces for this modular. We apply the modular to establish a pointwise selection principle, extending the classical Helly Selection Theorem.

5.1 The $\mathbb{N}$-Valued Pseudomodular

Let $I = [a, b] \subset \mathbb{R}$ be a closed interval with the end-points $a < b$, (M, d) be a metric space with metric d, and $X = M^I$ be the set of all mappings $x : I \to M$ from I into M. Given $x \in X$, the *oscillation* of x on I is the quantity

$$|x(I)| \equiv |x(I)|_d = \sup \{d(x(t), x(s)) : s, t \in I\},$$

also called the *diameter of the image* $x(I) = \{x(t) : t \in I\} \subset M$. A mapping x is said to be *bounded* if $|x(I)| < \infty$. We denote by $\mathrm{B}(I; M)$ the set of all bounded mappings from I into M. Given $x, y \in X$, the quantity $d_\infty(x, y) = \sup_{t \in I} d(x(t), y(t))$ is an extended metric on X and a metric on $\mathrm{B}(I; M)$, called the *uniform metric* (note that $d_\infty(x, y) \le d(x(a), y(a)) + |x(I)| + |y(I)|$).

If $n \in \mathbb{N}$, we denote by $\{I_i\}_1^n \prec I$ a collection of two-point subsets $I_i \equiv I_{s_i, t_i} = \{s_i, t_i\}$ of I ($i = 1, \ldots, n$) such that $s_1 < t_1 \le s_2 < t_2 \le \cdots \le s_{n-1} < t_{n-1} \le s_n < t_n$ (and so, the intervals $[s_1, t_1], \ldots, [s_n, t_n]$ are non-overlapping). Let $x, y \in X$. The value $|x(I_i)| = d(x(t_i), x(s_i))$ is just the increment of x on I_i. We define the *joint increment* of $x, y \in X$ on I_i by

$$|(x, y)(I_i)| = \sup_{v \in M} \left| d(x(t_i), v) + d(y(s_i), v) - d(x(s_i), v) - d(y(t_i), v) \right|. \qquad (5.1.1)$$

This quantity is well-defined: in fact, the absolute value under the supremum sign in (5.1.1) is less than or equal to (for any $v \in M$)

$$|d(x(t_i), v) - d(x(s_i), v)| + |d(y(s_i), v) - d(y(t_i), v)|$$

$$\le d(x(t_i), x(s_i)) + d(y(s_i), y(t_i)) = |x(I_i)| + |y(I_i)|,$$

© Springer International Publishing Switzerland 2015

V.V. Chistyakov, *Metric Modular Spaces*, SpringerBriefs in Mathematics,
DOI 10.1007/978-3-319-25283-4_5

and so, $|(x, y)(I_i)| \leq |x(I_i)| + |y(I_i)|$. We define the *joint oscillation* of $x, y \in X$ on I by

$$|(x, y)(I)| = \sup_{s,t \in I} |(x, y)(I_{s,t})|, \quad \text{where} \quad I_{s,t} = \{s, t\}.$$

Clearly, $|(x, y)(I)| \leq |x(I)| + |y(I)|$.

If $(M, d, +)$ is a metric semigroup (see Appendix A.3), then, instead of (5.1.1), one may consider the quantity

$$|(x, y)(I_i)| = d(x(t_i) + y(s_i), x(s_i) + y(t_i)), \tag{5.1.2}$$

whose properties are similar to those of (5.1.1), established (above and) below. Furthermore, if $(M, \| \cdot \|)$ is a normed linear space (over $\mathbb{R}$ or $\mathbb{C}$), one may also set

$$|(x, y)(I_i)| = \|(x - y)(t_i) - (x - y)(s_i)\| = \|x(t_i) + y(s_i) - x(s_i) - y(t_i)\|. \tag{5.1.3}$$

In what follows, we present arguments mostly for (more general) quantity (5.1.1).

Definition 5.1.1. For $\lambda > 0$ and $x, y \in X$, we define the quantity $w_\lambda(x, y) \equiv w_\lambda^N(x, y, I)$ valued in $\{0\} \cup \mathbb{N} \cup \{\infty\}$ by the rule:

$$w_\lambda(x, y) = \sup \left\{ n \in \mathbb{N} : \min_{1 \leq i \leq n} |(x, y)(I_i)| > \lambda \text{ for some } \{I_i\}_1^n \prec I \right\} \tag{5.1.4}$$

with $\sup \varnothing = 0$. Replacing I in (5.1.4) by any nonempty set $T \subset I$, we obtain the quantity denoted by $w_\lambda(x, y, T) \equiv w_\lambda^N(x, y, T)$.

Clearly, if $\varnothing \neq T_1 \subset T_2 \subset I$, we have $w_\lambda(x, y, T_1) \leq w_\lambda(x, y, T_2)$. The two extreme values of $w_\lambda(x, y)$, zero and infinity, are characterized as follows:

$$w_\lambda(x, y) = 0 \quad \text{iff} \quad |(x, y)(I)| \leq \lambda,$$

and so, $|(x, y)(I)| = \inf\{\lambda > 0 : w_\lambda(x, y) = 0\}$ ($\inf \varnothing = \infty$), and

$$w_\lambda(x, y) = \infty \quad \Leftrightarrow \quad \forall n \in \mathbb{N} \ \exists \{I_i\}_1^n \prec I \text{ such that } \min_{1 \leq i \leq n} |(x, y)(I_i)| > \lambda.$$

Lemma 5.1.2. *The function w from (5.1.4) is a pseudomodular on X and, given $x_0 \in M$, a modular on $\dot{X} = \{x \in X : x(a) = x_0\}$, which is continuous from the right, nonstrict and nonconvex on X and $\dot{X}$.*

Proof. As usual, we assume that $\lambda, \mu > 0$ and $x, y, z \in X$.

(i) By (5.1.1), given $n \in \mathbb{N}$ and $\{I_i\}_1^n \prec I$, $|(x, x)(I_i)| = 0$ for all $i = 1, \ldots, n$, and so, $w_\lambda(x, x) = \sup \varnothing = 0$.

Now, let us show that if $w_\lambda(x, y) = 0$ for all $\lambda > 0$, then

$$t \mapsto d(x(t), y(t)) \text{ is a constant function on } I.$$

Set $I_{s,t} = \{s, t\} \subset I$. Definition (5.1.4) implies $|(x, y)(I_{s,t})| \le \lambda$ for all $\lambda > 0$, and so, $|(x, y)(I_{s,t})| = 0$. It follows from (5.1.1) that

$$d(x(t), v) + d(y(s), v) = d(x(s), v) + d(y(t), v) \quad \text{for all} \quad v \in M.$$

Setting first $v = y(t)$ and then $v = x(s)$, we get

$$d(x(t), y(t)) + d(y(s), y(t)) = d(x(s), y(t)) = d(x(t), x(s)) + d(y(s), x(s)).$$

Similarly, setting $v = y(s)$ and then $v = x(t)$, we find

$$d(x(s), y(s)) + d(y(t), y(s)) = d(x(t), y(s)) = d(x(s), x(t)) + d(y(t), x(t)).$$

Thus, $d(x(t), y(t)) = d(x(s), y(s))$ (and $d(x(t), x(s)) = d(y(t), y(s))$) for all $s, t \in I$.

If, in addition, $x, y \in \dot{X}$, then $d(x(t), y(t)) = d(x(a), y(a)) = d(x_0, x_0) = 0$, i.e., $x(t) = y(t)$ for all $t \in I$, and so, $x = y$.

(ii) Since $|(x, y)(I_i)| = |(y, x)(I_i)|$ in (5.1.1), we have $w_\lambda(x, y) = w_\lambda(y, x)$.

(iii) In order to obtain inequality (iii) in Definition 1.2.1, first we show that

$$|(x, y)(I_i)| \le |(x, z)(I_i)| + |(z, y)(I_i)| \quad \text{for} \quad I_i = \{s_i, t_i\} \subset I. \tag{5.1.5}$$

In fact, if $v \in M$, the quantity under the supremum sign in (5.1.1) is estimated by

$$|d(x(t_i), v) + d(z(s_i), v) - d(x(s_i), v) - d(z(t_i), v)|$$
$$+ |d(z(t_i), v) + d(y(s_i), v) - d(z(s_i), v) - d(y(t_i), v)|$$
$$\le |(x, z)(I_i)| + |(z, y)(I_i)|,$$

and inequality (5.1.5) follows by taking the supremum over all $v \in M$.

Now, we prove that $w_{\lambda+\mu}(x, y) \le w_\lambda(x, z) + w_\mu(z, y)$. We may assume that $w_{\lambda+\mu}(x, y) \ne 0$. Let $n \in \mathbb{N}$, and $\{I_i\}_1^n \prec I$ be such that $|(x, y)(I_i)| > \lambda + \mu$ for all $1 \le i \le n$. If $1 \le i \le n$, (5.1.5) implies $|(x, z)(I_i)| > \lambda$ or $|(z, y)(I_i)| > \mu$, and so,

$$\{1, \dots, n\} = \{1 \le i \le n : |(x, z)(I_i)| > \lambda\} \cup \{1 \le i \le n : |(z, y)(I_i)| > \mu\}.$$

If $n_1 \ge 0$ and $n_2 \ge 0$ designate the numbers of elements in the first and second sets on the right, then (5.1.4) implies $n \le n_1 + n_2 \le w_\lambda(x, z) + w_\mu(z, y)$. The arbitrariness of n as above yields the inequality in axiom (iii).

To see that w is nonstrict, let $x, y \in X$ be bounded mappings. For any number $\lambda > |x(I)| + |y(I)|$, we find $\lambda > |(x, y)(I)|$, and so, $w_\lambda(x, y) = \sup \varnothing = 0$.

Let us show that w is nonconvex. Was it not so, inequality (iv) in Definition 1.2.1 would hold in particular with $\mu = \lambda > 0$ and $z = y$, i.e., $w_{2\lambda}(x, y) \leq w_\lambda(x, y)/2$. Let $x_0, x_1, y_0 \in M$, $x_0 \neq x_1$, and mappings $x, y \in X$ be given by: $x(t) = x_0$ if $a \leq t < b$, $x(b) = x_1$, and $y(t) = y_0$ for all $a \leq t \leq b$. Since

$$|(x, y)(I_{s,t})| = \sup_{v \in M} |d(x(t), v) - d(x(s), v)| = d(x(t), x(s)) \tag{5.1.6}$$

for any $I_{s,t} = \{s, t\} \subset I$, we find

$$w_\lambda(x, y) = \begin{cases} 1 & \text{if } 0 < \lambda < d(x_0, x_1), \\ 0 & \text{if } \qquad \lambda \geq d(x_0, x_1). \end{cases} \tag{5.1.7}$$

Now, if $0 < 2\lambda < d(x_0, x_1)$, we get $w_{2\lambda}(x, y) = 1 = w_\lambda(x, y)$, and so, inequality $w_{2\lambda}(x, y) \leq w_\lambda(x, y)/2$ does not hold.

Finally, we prove that w is continuous from the right. From (1.2.4), we know that $w_{\lambda+0}(x, y) \leq w_\lambda(x, y)$. To prove the reverse inequality, we may assume that $w_\lambda(x, y) \neq 0$. Let $n \in \mathbb{N}$ and $\{I_i\}_1^n \prec I$ be such that $|(x, y)(I_i)| > \lambda$ for all $i = 1, \ldots, n$. Choose λ' such that $\min_{1 \leq i \leq n} |(x, y)(I_i)| > \lambda' > \lambda$. It follows from (5.1.4) and (1.2.4) that $n \leq w_{\lambda'}(x, y) \leq w_{\lambda+0}(x, y)$. The definition of $w_\lambda(x, y)$ and the arbitrariness of n yield $w_\lambda(x, y) \leq w_{\lambda+0}(x, y)$. (Note that $\lambda \mapsto w_\lambda(x, y)$ is locally constant from the right, and, as (5.1.7) shows, w is not continuous from the left.) $\square$

Now we study the modular spaces X_w^*, X_w^0, and X_w^{fin} around a constant mapping $x^\circ(t) = x^\circ$ for all $t \in I$, where $x^\circ \in M$. Note that, by (5.1.6), the values $|(x, x^\circ)(I_i)|$ from (5.1.1) and $w_\lambda(x, x^\circ)$ from (5.1.4) are independent of x°.

Denote by $\text{Reg}(I; M)$ the set of all *regulated* mappings $x : I \to M$, i.e., satisfying the (right and left) *Cauchy condition* at every point of I: $d(x(t), x(s)) \to 0$ as $t, s \to \tau + 0$ for each $a \leq \tau < b$, and $d(x(t), x(s)) \to 0$ as $t, s \to \tau - 0$ for each $a < \tau \leq b$. If (M, d) is a *complete* metric space, then, by virtue of the *Cauchy criterion*, $x \in \text{Reg}(I; M)$ if and only if the right limit $x(\tau + 0) \in M$ exists at every point $a \leq \tau < b$ (i.e., $d(x(t), x(\tau + 0)) \to 0$ as $t \to \tau + 0$) and the left limit $x(\tau - 0) \in M$ exists at every point $a < \tau \leq b$ (and so, $d(x(t), x(\tau - 0)) \to 0$ as $t \to \tau - 0$).

Theorem 5.1.3. $X_w^{\text{fin}} = \text{Reg}(I; M) \subset X_w^* = X_w^0 = \text{B}(I; M)$.

In order to prove this theorem, we need a lemma. Recall that a sequence of mappings $\{x_n\} \subset X$ *converges pointwise on I to a mapping* $x \in X$ (in symbols: $x_n \to x$ pointwise on I) if $\lim_{n \to \infty} d(x_n(t), x(t)) = 0$ for all $t \in I$. Clearly, if $\{x_n\}$ *converges uniformly* to x, i.e., $\lim_{n \to \infty} d_\infty(x_n, x) = 0$, then $\{x_n\}$ converges pointwise to x (but not vice versa).

Lemma 5.1.4. (a) *Given $x, y \in X$, $a \leq s < t \leq b$, and $\lambda > 0$, we have: the quantity $n_t \equiv w_\lambda(x, y, [a, t])$ is finite if and only if $n_s \equiv w_\lambda(x, y, [a, s])$ and $n_{s,t} \equiv w_\lambda(x, y, [s, t])$ are finite, in which case we find $n_t = n_s + n_{s,t} + n_*$ with $n_* = 0$ or $n_* = 1$.*

(b) *If sequences $\{x_n\}, \{y_n\} \subset X$ converge pointwise on I to mappings $x, y \in X$, respectively, then $w_\lambda(x, y) \le \liminf_{n \to \infty} w_\lambda(x_n, y_n)$ for all $\lambda > 0$.*

Proof. (a) Since $n_s \le n_t$ and $n_{s,t} \le n_t$, we may suppose $n_t \ne 0$.

First, we show that if $n_t < \infty$, then $n_s + n_{s,t} \le n_t$. Assume that $n_s \ne 0$ and $n_{s,t} \ne 0$ (otherwise, the inequality is clear). By (5.1.4) with I replaced by $[a, s]$ and $[s, t]$, there are collections $\{I_i\}_1^{n_s} \prec [a, s]$ and $\{J_j\}_1^{n_{s,t}} \prec [s, t]$ such that $\min_{1 \le i \le n_s} |(x, y)(I_i)| > \lambda$ and $\min_{1 \le n \le n_{s,t}} |(x, y)(J_j)| > \lambda$. Since $\{I_i\}_1^{n_s} \cup \{J_j\}_1^{n_{s,t}} \prec [a, t]$ and $|(x, y)(I_i)| > \lambda$ for all $i = 1, \dots, n_s$ and $|(x, y)(J_j)| > \lambda$ for all $j = 1, \dots, n_{s,t}$, we get $n_s + n_{s,t} \le n_t$.

Second, assume that n_s and $n_{s,t}$ are finite, and let us prove that if $n \in \mathbb{N}$ and a collection $\{I_i\}_1^n \prec [a, t]$ with $I_i = \{s_i, t_i\}$ is such that $\min_{1 \le i \le n} |(x, y)(I_i)| > \lambda$, then $n \le n_s + n_{s,t} + 1$. If this is done, Definition (5.1.4) with $I = [a, t]$ and the arbitrariness of n as above imply $n_t \le n_s + n_{s,t} + 1$, which establishes the *equality* with $n_* \in \{0, 1\}$. We consider three possibilities. (I) If $\{I_i\}_1^n \prec [a, s]$, then $n \le n_s$, and similarly, if $\{I_i\}_1^n \prec [s, t]$, then $n \le n_{s,t}$. (II) If $n \ge 2$ and $s = t_j = s_{j+1}$ for some $1 \le j \le n - 1$, then $\{I_i\}_1^j \prec [a, s]$ and $\{I_i\}_{j+1}^n \prec [s, t]$, which implies $j \le n_s$ and $n - j \le n_{s,t}$, and so, $n \le n_s + n_{s,t}$. (III) Suppose $s_j < s < t_j$ for some $1 \le j \le n$. Clearly, if $n = 1$, then $n \le n_s + n_{s,t} + 1$, and if $n \ge 2$ then we have $\{I_i\}_1^{j-1} \prec [a, s]$ and $\{I_i\}_{j+1}^n \prec [s, t]$ (with $\{I_i\}_1^0 = \varnothing = \{I_i\}_{n+1}^n$). Therefore, (5.1.4) implies $j - 1 \le n_s$ and $n - j \le n_{s,t}$, and so, $n \le n_s + n_{s,t} + 1$.

(b) We may suppose $w_\lambda(x, y) \ne 0$. Let us show that if $w_\lambda(x, y)$ is finite and $k = w_\lambda(x, y)$, or $w_\lambda(x, y) = \infty$ and $k \in \mathbb{N}$ is arbitrary, then $k \le \liminf_{n \to \infty} w_\lambda(x_n, y_n)$. By Definition 5.1.1, there is a collection $\{I_i\}_1^k \prec I$ with $I_i = \{s_i, t_i\}$ such that $\min_{1 \le i \le k} |(x, y)(I_i)| > \lambda$. Let $\lambda' > 0$ be such that $\min_{1 \le i \le k} |(x, y)(I_i)| > \lambda' > \lambda$. Since $x_n \to x$ and $y_n \to y$ pointwise on I, there exists $n_0 \in \mathbb{N}$ (depending on $\{I_i\}_1^k$ and λ') such that the quantities $d(x_n(t_i), x(t_i))$, $d(x_n(s_i), x(s_i))$, $d(y_n(t_i), y(t_i))$, and $d(y_n(s_i), y(s_i))$ do not exceed $(\lambda' - \lambda)/4$ for all $1 \le i \le k$ and $n \ge n_0$. Noting that, by (5.1.1),

$$|(x, x_n)(I_i)| \le d(x(t_i), x_n(t_i)) + d(x(s_i), x_n(s_i)), \tag{5.1.8}$$

and applying inequality (5.1.5), we get

$$\lambda' < |(x, y)(I_i)| \le |(x, x_n)(I_i)| + |(x_n, y_n)(I_i)| + |(y_n, y)(I_i)|$$
$$\le d(x(t_i), x_n(t_i)) + d(x(s_i), x_n(s_i)) + |(x_n, y_n)(I_i)|$$
$$+ d(y_n(t_i), y(t_i)) + d(y_n(s_i), y(s_i))$$
$$\le |(x_n, y_n)(I_i)| + \lambda' - \lambda \quad \text{for all } 1 \le i \le k \text{ and } n \ge n_0.$$

Thus, $\min_{1 \le i \le k} |(x_n, y_n)(I_i)| > \lambda$ for all $n \ge n_0$. It follows from definition (5.1.4) that $k \le w_\lambda(x_n, y_n)$ for all $n \ge n_0$, and so,

$$k \le \inf_{n \ge n_0} w_\lambda(x_n, y_n) \le \liminf_{n \to \infty} w_\lambda(x_n, y_n). \qquad \square$$

Example 5.1.5. Here we illustrate that condition $n_* \in \{0, 1\}$ in Lemma 5.1.4(a) is essential. Let $x_0, x_1, x_2, y_0 \in M$ be such that $d(x_0, x_1) = d(x_1, x_2) = 1$ and $d(x_0, x_2) = 2$. Given $a < s < t = b$, define $x, y \in X$ by: $x(\tau) = x_0$ if $a \le \tau < s$, $x(s) = x_1, x(\tau) = x_2$ if $s < \tau \le b$, and $y(\tau) = y_0$ for all $\tau \in [a, b]$. Then n_s and $n_{s,t}$ are given as in (5.1.7), and so, $n_s = n_{s,t} = 1$ if $0 < \lambda < 1$, and $n_s = n_{s,t} = 0$ if $\lambda \ge 1$. Furthermore, $n_t = w_\lambda(x, y, [a, b]) = 2$ if $0 < \lambda < 1$, $n_t = 1$ if $1 \le \lambda < 2$, and $n_t = 0$ if $\lambda \ge 2$. Thus, $n_* = 0$ if $0 < \lambda < 1$ or $\lambda \ge 2$, and $n_* = 1$ if $1 \le \lambda < 2$.

Proof (of Theorem 5.1.3).

1. The inclusion $B(I; M) \subset X_w^0$ is clear: if $x \in B(I; M)$, then $|(x, x^\circ)(I)| = |x(I)| < \infty$, and so, $w_\lambda(x, x^\circ) = 0$ for all $\lambda > |(x, x^\circ)(I)|$, which implies $x \in X_w^0 \subset X_w^*$.
2. In order to see that $X_w^* \subset B(I; M)$, we show that

   $$\text{if } x \in X \text{ and } |x(I)| = \infty, \text{ then } w_\lambda(x, x^\circ) = \infty \text{ for all } \lambda > 0.$$

   Let $\lambda > 0$. Condition $|x(I)| = \infty$ is equivalent to $\sup_{t \in I} d(x(t), x(s)) = \infty$ for all $s \in I$. Set $s_0 = a$, and pick $s_1 \in I$ such that $d(x(s_1), x(s_0)) > \lambda$. Inductively, if $k \ge 2$ and numbers $s_0, s_1, \ldots, s_{k-1} \in I$ are already chosen, we pick $s_k \in I$ such that

   $$d(x(s_k), x(s_{k-1})) > \lambda + \sum_{i=1}^{k-1} d(x(s_i), x(s_{i-1})). \tag{5.1.9}$$

   Given $n \in \mathbb{N}$, we re-order the collection of pairwise different points $s_0, s_1, \ldots, s_n$ in ascending order and denote them by $t_0 < t_1 < \cdots < t_n$. Setting $I_i = \{t_{i-1}, t_i\}$ for $i = 1, \ldots, n$, we find $\{I_i\}_1^n \prec I$. Since $|(x, x^\circ)(I_i)| = |x(I_i)|$, if we show that $|x(I_i)| > \lambda$ for every $i \in \{1, \ldots, n\}$, then the arbitrariness of n and (5.1.4) will imply $w_\lambda(x, x^\circ) = \infty$. So, let $1 \le i \le n$ and $I_i = \{s_k, s_m\}$ for some $0 \le k, m \le n$. We have $|x(I_i)| = d(x(s_k), x(s_m))$, and we may assume that $m < k$. If $m = k - 1$, then $|x(I_i)| > \lambda$ thanks to (5.1.9). For $m \le k - 2$, the triangle inequality for d yields

   $$d(x(s_k), x(s_{k-1})) \le d(x(s_k), x(s_m)) + \sum_{i=m+1}^{k-1} d(x(s_{i-1}), x(s_i))$$

   $$\le |x(I_i)| + \sum_{i=1}^{k-1} d(x(s_{i-1}), x(s_i)),$$

   which together with (5.1.9) proves that $|x(I_i)| > \lambda$.
3. Let us prove that $X_w^{\text{fin}} \subset \text{Reg}(I; M)$. Suppose $x \in X_w^{\text{fin}}$, so that $w_\lambda(x, x^\circ) < \infty$ for all $\lambda > 0$. Given $a \le \tau < b$, let us verify the (right) Cauchy condition at τ: for every $\varepsilon > 0$ there is $0 < \delta(\varepsilon) < b - \tau$ such that $d(x(t), x(s)) \le \varepsilon$ for all $\tau < s, t \le \tau + \delta(\varepsilon)$. On the contrary, assume that there exists $\varepsilon_0 > 0$ such that

if $0 < \delta < b - \tau$, then there are s_δ and t_δ such that $\tau < s_\delta < t_\delta \leq \tau + \delta$ and $d(x(t_\delta), x(s_\delta)) > \varepsilon_0$. Given $0 < \delta_1 < b - \tau$, pick s_1 and t_1 such that $\tau < s_1 < t_1 \leq \tau + \delta_1$ and $d(x(t_1), x(s_1)) > \varepsilon_0$. Inductively, if $i \geq 2$, and $0 < \delta_{i-1} < b - \tau$ and two points s_{i-1} and t_{i-1} satisfying $\tau < s_{i-1} < t_{i-1} \leq \tau + \delta_{i-1}$ are already chosen, we set $\delta_i = s_{i-1} - \tau$ and pick points s_i and t_i such that $\tau < s_i < t_i \leq \tau + \delta_i = s_{i-1}$ and $d(x(t_i), x(s_i)) > \varepsilon_0$. Let $n \in \mathbb{N}$, and $I_i = \{s_{n-i+1}, t_{n-i+1}\}$, $i = 1, \ldots, n$. By the construction, $\{I_i\}_1^n \prec (\tau, b) \subset [a, b]$ and $|(x, x^\circ)(I_i)| = |x(I_i)| > \varepsilon_0$ for all $1 \leq i \leq n$. Since n is arbitrary, (5.1.4) implies $w_\lambda(x, x^\circ) = \infty$ with $\lambda = \varepsilon_0$, which is a contradiction.

The arguments for $a < \tau \leq b$ and the left Cauchy condition are similar.

4. Now, we show that $\mathrm{Reg}(I; M) \subset X_w^{\mathrm{fin}}$. Suppose x is a regulated mapping and, for contradiction, $w_{\lambda_0}(x, x^\circ, [a, b]) = \infty$ for some $\lambda_0 > 0$. Set $[a_0, b_0] = [a, b]$ and $c_0 = (a_0 + b_0)/2$. By Lemma 5.1.4(a) with $y = x^\circ$, at least one of the quantities $w_{\lambda_0}(x, x^\circ, [a_0, c_0])$ or $w_{\lambda_0}(x, x^\circ, [c_0, b_0])$ is infinite. We denote by $[a_1, b_1]$ any of the intervals $[a_0, c_0]$ or $[c_0, b_0]$, for which $w_{\lambda_0}(x, x^\circ, [a_1, b_1]) = \infty$. Inductively, if $k \geq 2$ and an interval $[a_{k-1}, b_{k-1}] \subset [a, b]$ such that $w_{\lambda_0}(x, x^\circ, [a_{k-1}, b_{k-1}]) = \infty$ is already chosen, we denote by c_{k-1} the middle point of $[a_{k-1}, b_{k-1}]$ and by $[a_k, b_k]$ one of the intervals $[a_{k-1}, c_{k-1}]$ or $[c_{k-1}, b_{k-1}]$, for which $w_{\lambda_0}(x, x^\circ, [a_k, b_k]) = \infty$. In this way, for every $k \in \mathbb{N}$ we obtain nested intervals $[a_k, b_k] \subset [a_{k-1}, b_{k-1}]$ from $[a, b]$ such that $w_{\lambda_0}(x, x^\circ, [a_k, b_k]) = \infty$ and $b_k - a_k = (b - a)/2^k$. Let $\tau \in [a, b]$ be the common point of all intervals $[a_k, b_k]$, so that $a_k \to \tau$ and $b_k \to \tau$ as $k \to \infty$. Assume that $a < \tau < b$ (the cases $\tau = a$ and $\tau = b$ are considered similarly). Since $\tau \in [a_k, b_k]$ and $w_{\lambda_0}(x, x^\circ, [a_k, b_k]) = \infty$ for all $k \in \mathbb{N}$, Lemma 5.1.4(a) with $y = x^\circ$ implies the existence of a subsequence $\{k_l\}_{l=1}^\infty$ of $\{k\}_{k=1}^\infty$ such that $w_{\lambda_0}(x, x^\circ, [a_{k_l}, \tau]) = \infty$ for all $l \in \mathbb{N}$, or $w_{\lambda_0}(x, x^\circ, [\tau, b_{k_l}]) = \infty$ for all $l \in \mathbb{N}$. To be specific, suppose the latter possibility takes place. Clearly, $\tau < b_{k_l}$ for all $l \in \mathbb{N}$. For every $0 < \delta < b - \tau$ choose $l(\delta) \in \mathbb{N}$ such that $\tau < b_{k_{l(\delta)}} \leq \tau + \delta$. Since $[\tau, b_{k_{l(\delta)}}] \subset [\tau, \tau + \delta]$, we find $w_{\lambda_0}(x, x^\circ, [\tau, \tau + \delta]) = \infty$. This implies $w_{\lambda_0}(x, x^\circ, (\tau, \tau + \delta]) = \infty$ (see the next paragraph), and so, by (5.1.4) with I replaced by $(\tau, \tau + \delta]$, there exist s and t such that $\tau < s < t \leq \tau + \delta$ and $d(x(t), x(s)) > \lambda_0$, which contradicts the (right) Cauchy condition at the point τ.

To see that $w_{\lambda_0}(x, x^\circ, (\tau, \tau + \delta]) = \infty$, we note that, for every $n \in \mathbb{N}$, condition $w_{\lambda_0}(x, x^\circ, [\tau, \tau + \delta]) = \infty$ yields the existence of $\{I_i\}_1^n \prec [\tau, \tau + \delta]$ such that $\min_{1 \leq i \leq n} |x(I_i)| > \lambda_0$. It follows that $\{I_i\}_2^n \prec (\tau, \tau + \delta]$ and $\min_{2 \leq i \leq n} |x(I_i)| > \lambda_0$. Hence $n - 1 \leq w_{\lambda_0}(x, x^\circ, (\tau, \tau + \delta])$ for every $n \in \mathbb{N}$. $\qquad \square$

Example 5.1.6. Let $\mathscr{D} : [a, b] \to M$ be a Dirichlet-type mapping defined by the rule: $\mathscr{D}(t) = x_0$ if $t \in [a, b] \cap \mathbb{Q}$, and $\mathscr{D}(t) = x_1$ if $t \in [a, b] \setminus \mathbb{Q}$, where $x_0, x_1 \in M$, $x_0 \neq x_1$. By Theorem 5.1.3, $\mathscr{D} \in X_w^* = X_w^0$ and $\mathscr{D} \notin X_w^{\mathrm{fin}}$. Furthermore,

$$w_\lambda(\mathscr{D}, x^\circ, [a, b]) = \begin{cases} \infty & \text{if } 0 < \lambda < d(x_0, x_1), \\ 0 & \text{if } \quad \lambda \geq d(x_0, x_1). \end{cases}$$

By Lemma 5.1.2, w from (5.1.4) is a pseudomodular on X, and so, Theorems 2.2.1 and 5.1.3 imply that $d_w^0(x, y) = \inf\{\lambda > 0 : w_\lambda(x, y) \le \lambda\}$ is an extended pseudometric on X and a pseudometric on $X_w^* = \mathrm{B}(I; M)$. We are going to study the relationship between d_w^0 and the uniform metric d_∞ on $\mathrm{B}(I; M)$.

Lemma 5.1.7. *Given* $x, y \in \mathrm{B}(I; M)$, *we have:*

(a) $d_w^0(x, y) \le |(x, y)(I)| \le 2d_\infty(x, y)$ *and* $d_\infty(x, y) \le d(x(a), y(a)) + |(x, y)(I)|$;

(b) *if* $d_w^0(x, y) < 1$, *then* $d_w^0(x, y) = |(x, y)(I)|$, *and so*, $d_\infty(x, y) \le d^1(x, y)$, *where* $d^1(x, y) = d(x(a), y(a)) + d_w^0(x, y)$ *is a metric on* $\mathrm{B}(I; M)$.

Proof. (a) Since $w_\lambda(x, y) = 0 < \lambda$ for all $\lambda > |(x, y)(I)|$, $d_w^0(x, y) \le \lambda$, and so, $d_w^0(x, y) \le |(x, y)(I)|$. This inequality may be strict: in the context of example (5.1.7), we have $|(x, y)(I)| = d(x_0, x_1)$, $d_w^0(x, y) = d(x_0, x_1)$ if $d(x_0, x_1) \le 1$, and $d_w^0(x, y) = 1$ if $d(x_0, x_1) > 1$.

Now, if $I_{s,t} = \{s, t\} \subset I$, then (cf. (5.1.8))

$$|(x, y)(I_{s,t})| \le d(x(t), y(t)) + d(x(s), y(s)) \le 2d_\infty(x, y),$$

and so, $|(x, y)(I)| \le 2d_\infty(x, y)$. On the other hand, given $v \in M$,

$$|d(x(t), v) - d(y(t), v)| \le |d(x(t), v) + d(y(s), v) - d(x(s), v) - d(y(t), v)|$$

$$+ |d(x(s), v) - d(y(s), v)|$$

$$\le |(x, y)(I_{s,t})| + d(x(s), y(s)),$$

whence, taking the supremum over all $v \in M$,

$$d(x(t), y(t)) \le |(x, y)(I_{s,t})| + d(x(s), y(s)) \quad \text{for all } s, t \in I. \qquad (5.1.10)$$

The second inequality in (a) follows readily with $s = a$.

(b) For any $\lambda > 0$ such that $d_w^0(x, y) < \lambda < 1$, we have $w_\lambda(x, y) \le \lambda < 1$, i.e., $w_\lambda(x, y) = 0$. This implies $|(x, y)(I)| \le \lambda$, and so, $|(x, y)(I)| \le d_w^0(x, y)$. The reverse inequality follows from the first inequality in (a).

Inequality $d_\infty(x, y) \le d^1(x, y)$ is a consequence of the third inequality in (a). This proves also that d^1 is a metric on $\mathrm{B}(I; M)$: in fact, d^1 is a pseudometric, and if $d^1(x, y) = 0$, then $d_\infty(x, y) = 0$, and so, $x = y$. $\qquad\square$

The last lemma implies that the d^1-convergence and uniform convergence in $\mathrm{B}(I; M)$ are equivalent: if $d_w^0(x, y) < 1$, then $d_\infty(x, y) \le d^1(x, y) \le 3d_\infty(x, y)$. In particular, convergences with respect to metrics d_w^0 and d_∞ are equivalent on the set

$$\dot{X}_w^* = X_w^* \cap \dot{X} = \{x \in \mathrm{B}(I; M) : x(a) = x_0\}.$$

Combining Theorem 4.1.1(c) and Lemma 5.1.7(a), we get:

if $\{x_n\} \subset \mathrm{Reg}(I; M)$, $x \in X$, and $d_\infty(x_n, x) \to 0$, then $x \in \mathrm{Reg}(I; M)$.

The modular convergence in $\dot{X}_w^*$ is characterized as follows.

Lemma 5.1.8. *Given a sequence $\{x_n\} \subset \dot{X}_w^*$ and $x \in \dot{X}_w^*$, we have: $x_n \overset{w}{\to} x$ if and only if $\limsup_{n \to \infty} d_\infty(x_n, x) < \infty$.*

Proof. ($\Rightarrow$) By the definition of the w-convergence, $w_{\lambda_0}(x_n, x) \to 0$ for some $\lambda_0 > 0$, and so, there is $n_0 \in \mathbb{N}$ such that $w_{\lambda_0}(x_n, x) < 1$ for all $n \geq n_0$. Hence $w_{\lambda_0}(x_n, x) = 0$ implying $|(x_n, x)(I)| \leq \lambda_0$. Since $x_n(a) = x(a) = x_0$, applying Lemma 5.1.7(a), we get $d_\infty(x_n, x) \leq |(x_n, x)(I)| \leq \lambda_0$ for all $n \geq n_0$. Thus,

$$\limsup_{n \to \infty} d_\infty(x_n, x) \leq \sup_{n \geq n_0} d_\infty(x_n, x) \leq \lambda_0 < \infty.$$

($\Leftarrow$) Suppose $\limsup_{n \to \infty} d_\infty(x_n, x) < \lambda_0 < \infty$. Then $\sup_{n \geq n_0} d_\infty(x_n, x) < \lambda_0$ for some $n_0 \in \mathbb{N}$, and so, by virtue of Lemma 5.1.7(a),

$$|(x_n, x)(I)| \leq 2d_\infty(x_n, x) < 2\lambda_0 \quad \text{for all } n \geq n_0.$$

It follows that $w_{2\lambda_0}(x_n, x) = 0$, $n \geq n_0$, which implies $x_n \overset{w}{\to} x$. $\quad\square$

5.2   The Pointwise Selection Principle

In this section, basing on the $\mathbb{N}$-valued pseudomodular $w = w^{\mathbb{N}}$ from (5.1.4) we present a *pointwise selection principle* for sequences of regulated or non-regulated mappings (Theorem 5.2.1). This is a far-reaching extension of the classical *Helly Selection Theorem*, which we recall now: *a uniformly bounded sequence (or infinite family) of monotone real functions on an interval $I = [a, b]$ contains a pointwise convergent subsequence whose pointwise limit is a bounded monotone function on I.*

A sequence of mappings $\{x_n\} \subset X = M^I$ is said to be *pointwise precompact* (on I) if the closure in M of the set $\{x_n(t) : n \in \mathbb{N}\}$ is compact for all $t \in I$.

Theorem 5.2.1. *Let $\{x_n\} \subset X = M^I$ be a pointwise precompact sequence, $\{y_n\} \subset X$ be a pointwise convergent sequence with the pointwise limit $y \in X$, and*

$$\mathbf{w}(\lambda) \equiv \limsup_{n \to \infty} w_\lambda(x_n, y_n) < \infty \quad \text{for all } \lambda > 0. \tag{5.2.1}$$

Then, there is a subsequence of $\{x_n\}$, which converges pointwise on I to a mapping $x \in X$ such that $w_\lambda(x, y) \leq \mathbf{w}(\lambda)$ for all $\lambda > 0$.

Proof. Denote by $\mathrm{Mon}(I; \mathbb{N})$ the set of all bounded nondecreasing functions, which map $I = [a, b]$ into $\mathbb{N}$.

Note that, for every $\lambda > 0$ and $n \in \mathbb{N}$, the function $t \mapsto f(\lambda, n, t) \equiv w_\lambda(x_n, y_n, [a, t])$ is nondecreasing on I, and $f(\lambda, n, t) \leq w_\lambda(x_n, y_n) \equiv w_\lambda(x_n, y_n, I)$ for all $t \in I$.

1. Making use of the diagonal process, let us show that, given a decreasing sequence $\{\lambda_k\}_{k=1}^\infty \subset (0, \infty)$ tending to zero, there exists an increasing sequence $\{n_j\}_{j=1}^\infty \subset \mathbb{N}$, and for every $k \in \mathbb{N}$ there exists a function $N_k \in \mathrm{Mon}(I; \mathbb{N})$ such that

$$\lim_{j \to \infty} w_{\lambda_k}(x_{n_j}, y_{n_j}, [a, t]) = N_k(t) \quad \text{for all } t \in I. \tag{5.2.2}$$

By assumption (5.2.1), for every $\lambda > 0$ there are $m_0(\lambda), n_0(\lambda) \in \mathbb{N}$ such that $w_\lambda(x_n, y_n) \leq m_0(\lambda)$ for all $n \geq n_0(\lambda)$. The sequence of functions

$$\{t \mapsto f(\lambda_1, n, t)\}_{n=n_0(\lambda_1)}^\infty \subset \mathrm{Mon}(I; \mathbb{N})$$

is uniformly bounded on I by constant $m_0(\lambda_1)$, and so, by the Helly Selection Theorem, there exists an increasing sequence $\ell_1 : \mathbb{N} \to \mathbb{N}$ with $\ell_1(1) \geq n_0(\lambda_1)$ and a function $N_1 \in \mathrm{Mon}(I; \mathbb{N})$ such that $f(\lambda_1, \ell_1(j), t)$ converges to $N_1(t)$ as $j \to \infty$ for all $t \in I$. Pick the least number $j_1 \in \mathbb{N}$ such that $\ell_1(j_1) \geq n_0(\lambda_2)$. Then, the sequence of functions $\{t \mapsto f(\lambda_2, \ell_1(j), t)\}_{j=j_1}^\infty \subset \mathrm{Mon}(I; \mathbb{N})$ is uniformly bounded on I by constant $m_0(\lambda_2)$. Applying Helly's Theorem, we find a subsequence $\{\ell_2(j)\}_{j=1}^\infty$ of $\{\ell_1(j)\}_{j=j_1}^\infty$ and a function $N_2 \in \mathrm{Mon}(I; \mathbb{N})$ such that $f(\lambda_2, \ell_2(j), t)$ converges to $N_2(t)$ as $j \to \infty$ for all $t \in I$. Choose the least number $j_2 \in \mathbb{N}$ such that $\ell_2(j_2) \geq n_0(\lambda_3)$. Inductively, if $k \geq 3$, and an increasing sequence $\ell_{k-1} : \mathbb{N} \to \mathbb{N}$ and a number $j_{k-1} \in \mathbb{N}$ such that $\ell_{k-1}(j_{k-1}) \geq n_0(\lambda_k)$ are already chosen, then, by the Helly Selection Theorem applied to the sequence of functions

$$\{t \mapsto f(\lambda_k, \ell_{k-1}(j), t)\}_{j=j_{k-1}}^\infty \subset \mathrm{Mon}(I; \mathbb{N}),$$

which is uniformly bounded by constant $m_0(\lambda_k)$, we find a subsequence $\{\ell_k(j)\}_{j=1}^\infty$ of $\{\ell_{k-1}(j)\}_{j=j_{k-1}}^\infty$ and a function $N_k \in \mathrm{Mon}(I; \mathbb{N})$ such that $f(\lambda_k, \ell_k(j), t)$ converges to $N_k(t)$ as $j \to \infty$ for all $t \in I$, i.e.,

$$\lim_{j \to \infty} w_{\lambda_k}(x_{\ell_k(j)}, y_{\ell_k(j)}, [a, t]) = N_k(t) \quad \text{for all } t \in I.$$

Now, given $k \in \mathbb{N}$, the sequence $\{\ell_j(j)\}_{j=k}^\infty$ is a subsequence of $\{\ell_k(j)\}_{j=1}^\infty$, and so, setting $n_j = \ell_j(j)$ for all $j \in \mathbb{N}$ (i.e., $\{n_j\}_{j=1}^\infty$ is the increasing diagonal sequence), we obtain (5.2.2).

For the sake of brevity, we write $x_j = x_{n_j}$ and $y_j = y_{n_j}$ in (5.2.2).

2. The set $S_k \subset I$ of discontinuity points of each function $N_k \in \text{Mon}(I; \mathbb{N})$ is at most countable. If S denotes the union of all rational numbers from I and $\bigcup_{k=1}^{\infty} S_k$, then S is an at most countable dense subset of I. Furthermore, we have:

$$N_k \text{ is continuous at all points of } I \setminus S \text{ for all } k \in \mathbb{N}. \qquad (5.2.3)$$

Since the set $\{x_j(t)\}_{j=1}^{\infty}$ is precompact in M for all $t \in I$, and $S \subset I$ is at most countable, we may assume (applying the diagonal procedure and passing to a subsequence of $\{x_j\}$ if necessary) that, for every $s \in S$, $x_j(s)$ converges in M as $j \to \infty$ to a point denoted by $x(s) \in M$.

Now, we show that, given $t \in I \setminus S$, the sequence $\{x_j(t)\}$ is convergent in M. Let $\varepsilon > 0$. Since $\lambda_k \to 0$ as $k \to \infty$, choose and fix $k = k(\varepsilon) \in \mathbb{N}$ such that $\lambda_k \leq \varepsilon$. By (5.2.3), N_k is continuous at t and, by the density of S in I, there is $s = s(t, k) \in S$ such that $|N_k(t) - N_k(s)| < 1$, i.e., $N_k(t) = N_k(s)$. Taking into account (5.2.2), let $J_1 = J_1(t, k)$ and $J_2 = J_2(s, k)$ be two positive integers such that if $j \geq \max\{J_1, J_2\}$,

$$w_{\lambda_k}(x_j, y_j, [a, t]) = N_k(t) \quad \text{and} \quad w_{\lambda_k}(x_j, y_j, [a, s]) = N_k(s).$$

Assuming that $s < t$ (with no loss of generality) and applying Lemma 5.1.4(a), we get, for all $j \geq \max\{J_1, J_2\}$,

$$w_{\lambda_k}(x_j, y_j, [s, t]) \leq w_{\lambda_k}(x_j, y_j, [a, t]) - w_{\lambda_k}(x_j, y_j, [a, s])$$
$$= N_k(t) - N_k(s) = 0.$$

Hence $w_{\lambda_k}(x_j, y_j, [s, t]) = 0$. If $I_{s,t} = \{s, t\}$, from (5.1.4) with I replaced by $[s, t]$, we find $|(x_j, y_j)(I_{s,t})| \leq \lambda_k \leq \varepsilon$ for all $j \geq \max\{J_1, J_2\}$. Being convergent, $\{x_j(s)\}$, $\{y_j(t)\}$ and $\{y_j(s)\}$ are Cauchy sequences in M, and so, there exists a positive integer $J_3 = J_3(\varepsilon, t, s)$ such that if $j, j' \geq J_3$, we have

$$d(x_j(s), x_{j'}(s)) \leq \varepsilon, \quad d(y_j(t), y_{j'}(t)) \leq \varepsilon, \quad \text{and} \quad d(y_j(s), y_{j'}(s)) \leq \varepsilon.$$

By the inequality similar to (5.1.8), we get

$$|(y_j, y_{j'})(I_{s,t})| \leq d(y_j(t), y_{j'}(t)) + d(y_j(s), y_{j'}(s)) \leq 2\varepsilon.$$

Noting that the number $J = \max\{J_1, J_2, J_3\}$ depends only on ε and applying inequalities (5.1.10) and (5.1.5), we find that if $j, j' \geq J$, then

$$d(x_j(t), x_{j'}(t)) \leq d(x_j(s), x_{j'}(s)) + |(x_j, x_{j'})(I_{s,t})|$$
$$\leq d(x_j(s), x_{j'}(s)) + |(x_j, y_j)(I_{s,t})| + |(y_j, y_{j'})(I_{s,t})| + |(y_{j'}, x_{j'})(I_{s,t})|$$
$$\leq \varepsilon + \varepsilon + 2\varepsilon + \varepsilon = 5\varepsilon.$$

This proves that $\{x_j(t)\}$ is a Cauchy sequence in M, and, since it is precompact, $x_j(t)$ converges in M as $j \to \infty$ to a point denoted by $x(t) \in M$. It follows that the mapping $x : I = S \cup (I \setminus S) \to M$ is well-defined, and it is the pointwise limit on I of the sequence $\{x_j\} = \{x_{n_j}\}$ (which is a subsequence of the original sequence $\{x_n\}$). Noting that $\{y_j\} = \{y_{n_j}\}$ converges pointwise on I to the mapping y as $j \to \infty$ and applying Lemma 5.1.4(b), we get, for all $\lambda > 0$,

$$w_\lambda(x, y) \le \liminf_{j \to \infty} w_\lambda(x_{n_j}, y_{n_j}) \le \limsup_{n \to \infty} w_\lambda(x_n, y_n) = \mathbf{w}(\lambda).$$

This completes the proof of Theorem 5.2.1.   □

Remark 5.2.2. In Theorem 5.2.1, a pointwise convergent subsequence of $\{x_n\}$ is extracted with respect to a given convergent sequence $\{y_n\}$ (or a mapping $y \in X$ if $y_n = y$ for all $n \in \mathbb{N}$). Traditionally (e.g., in Helly's Theorem), the sequence $\{y_n\}$ is constant-valued in that, for some $y_0 \in M$, $y_n(t) = y_0$ for all $t \in I$ and $n \in \mathbb{N}$. The conclusion '$w_\lambda(x, y) \le \mathbf{w}(\lambda)$ for all $\lambda > 0$' means a certain *regularity* of the pointwise limit $x \in X$ in the sense that $x \in X_w^{\text{fin}}(y)$. In particular, if $y = x^\circ$, then, by Theorem 5.1.3, we get $x \in X_w^{\text{fin}}(x^\circ) = \text{Reg}(I; M)$.

Example 5.2.3. (1) Condition (5.2.1) is sufficient, but not necessary. In fact, let I be the unit interval $[0, 1]$, $x_0, x_1 \in M$, $x_0 \ne x_1$, and, given $n \in \mathbb{N}$ and $t \in I$, define $x_n(t) \in M$ by: $x_n(t) = x_0$ if $n!t \in \mathbb{Z}$, and $x_n(t) = x_1$ otherwise. The sequence of regulated mappings $\{x_n\} \subset X$ converges pointwise on I to the Dirichlet mapping $\mathscr{D}$ from Example 5.1.6, and

$$w_\lambda(x_n, x^\circ, [0, 1]) = \begin{cases} 2 \cdot n! & \text{if } 0 < \lambda < d(x_0, x_1), \\ 0 & \text{if } \quad \lambda \ge d(x_0, x_1). \end{cases}$$

(2) The choice of an appropriate sequence $\{y_n\}$ in Theorem 5.2.1 is essential. Let $\{x_{0n}\}, \{x_{1n}\} \subset M$ be two sequences such that $x_{0n} \to x_0$ and $x_{1n} \to x_1$ as $n \to \infty$, where $x_0, x_1 \in M$, $x_0 \ne x_1$. Given $n \in \mathbb{N}$, define $x_n : I \to M$ by: $x_n(t) = x_{0n}$ if $t \in I \cap \mathbb{Q}$, and $x_n(t) = x_{1n}$ if $t \in I \setminus \mathbb{Q}$. So (cf. Example 5.1.6),

$$w_\lambda(x_n, x^\circ) = \begin{cases} \infty & \text{if } 0 < \lambda < d(x_{0n}, x_{1n}), \\ 0 & \text{if } \quad \lambda \ge d(x_{0n}, x_{1n}), \end{cases}$$

and $\{x_n\}$ converges pointwise on I to the Dirichlet mapping $\mathscr{D}$. Since

$$|d(x_{0n}, x_{1n}) - d(x_0, x_1)| \le d(x_{0n}, x_0) + d(x_{1n}, x_1) \to 0 \quad \text{as } n \to \infty,$$

there is $n_0 \in \mathbb{N}$ such that $d(x_{0n}, x_{1n}) > d(x_0, x_1)/2 \equiv \lambda_0$ for all $n \ge n_0$. Hence $w_\lambda(x_n, x^\circ) = \infty$ for all $0 < \lambda \le \lambda_0$ and $n \ge n_0$, and so, Theorem 5.2.1 is inapplicable in this context.

On the other hand, if $y_n = \mathscr{D}$, $n \in \mathbb{N}$, and $I_{s,t} = \{s, t\}$, we have (cf. (5.1.8))

$$|(x_n, \mathscr{D})(I_{s,t})| \leq d(x_n(t), \mathscr{D}(t)) + d(x_n(s), \mathscr{D}(s)) \leq 2\varepsilon_n,$$

where $\varepsilon_n = \max\{d(x_{0n}, x_0), d(x_{1n}, x_1)\}$. Now, if $\lambda > 0$, there is $n_0 = n_0(\lambda) \in \mathbb{N}$ such that $2\varepsilon_n \leq \lambda$ for all $n \geq n_0$, and so, by (5.1.4), $w_\lambda(x_n, \mathscr{D}) = 0$, which implies condition (5.2.1). It remains to note that the sequence $\{x_n\}$ is pointwise precompact.

5.3 Bibliographical Notes and Comments

Section 5.1. It was discovered in this section that w from (5.1.4) has modular properties. The restricted quantity $w_\lambda(x, x^\circ)$ when x° is a constant mapping and $M = \mathbb{R}$ seems to originate from Dudley and Norvaiša [37, Part III, Sect. 2]. Theorem 5.1.3 and Lemma 5.1.4 were established in Chistyakov, Maniscalco and Tretyachenko [31, Sect. 3] (in different terminology) along with many other properties. Examples for real-valued functions are presented in Tret'yachenko and Chistyakov [101].

Section 5.2. The first pointwise selection principle for monotone functions is due to Helly [44]; its exposition can be found in many textbooks, e.g., Kelley [52], Kolmogorov and Fomin [54], Natanson [83]. The novelty in Theorem 5.2.1, which is an extension of [31, Theorem 2.1], is the introduction of an appropriate pointwise convergent sequence $\{y_n\} \subset X$. At the same time, this Theorem contains as particular cases many pointwise selection theorems established for different classes of functions and mappings in Belov and Chistyakov [7], Chistyakov [10–12, 14–16, 18, 21, 31], Dudley and Norvaiša [37], Gniłka [40], Musielak and Orlicz [76], Schramm [96], Waterman [103]. More details can be found in [31].

Chapter 6
Mappings of Bounded Generalized Variation

Abstract Here we follow the notation of Chap. 5, and denote the pseudomodular from Definition 5.1.1 more precisely by $w_\lambda^N(x, y)$. For $I = [a, b]$ and a metric space (M, d), we define new pseudomodulars on the set $X = M^I$, whose induced modular spaces consist of mappings of bounded generalized variation (in the sense of Jordan, Wiener-Young, Riesz-Medvedev). We prove the Lipschitz continuity of a superposition operator (of "multiplication") and establish the existence of selections of bounded variation of compact-valued BV multifunctions. An application to ordinary differential equations in Banach spaces is also given.

6.1 The Wiener-Young Variation

Let $\varphi : [0, \infty) \to [0, \infty)$ be a nondecreasing continuous function such that $\varphi(0) = 0$, $\varphi(u) > 0$ for $u > 0$, and $\varphi(\infty) = \infty$ (i.e., $\lim_{u \to \infty} \varphi(u) = \infty$). Such functions are conventionally called φ-functions.

Definition 6.1.1. Given $\lambda > 0$ and $x, y \in X$, we define $w : (0, \infty) \times X \times X \to [0, \infty]$ by

$$w_\lambda(x, y) \equiv w_\lambda^\varphi(x, y, I) = \sup\left\{ \sum_{i=1}^n \varphi\left(\frac{|(x, y)(I_i)|}{\lambda} \right) : n \in \mathbb{N} \text{ and } \{t_i\}_0^n \prec I \right\}, \quad (6.1.1)$$

where $\{t_i\}_0^n \prec I$ denotes a *partition* of I, i.e., $a = t_0 < t_1 < \cdots < t_{n-1} < t_n = b$, and $|(x, y)(I_i)|$ is given by (5.1.1) (or (5.1.2), or (5.1.3)) with $I_i = \{t_{i-1}, t_i\}$, $i = 1, \ldots, n$. (Note that $\{I_i\}_1^n \prec I$ with $I_i = \{s_i, t_i\}$ is a partition of I iff $\sum_{i=1}^n |I_i| = |I| \equiv b - a$, where $|I_i| = t_i - s_i$.) The expression $\varphi(|(x, y)(I_i)|/\lambda)$ under the summation sign in (6.1.1) is called the *variational core* of w corresponding to $\{t_i\}_0^n \prec I$.

Lemma 6.1.2. *The function w from (6.1.1) is a pseudomodular on X, and given $x_0 \in M$, a strict modular on $\dot{X} = \{x \in X : x(a) = x_0\}$. Moreover, w is continuous from the right, and if φ is convex on $[0, \infty)$, then w is convex.*

© Springer International Publishing Switzerland 2015
V.V. Chistyakov, *Metric Modular Spaces*, SpringerBriefs in Mathematics,
DOI 10.1007/978-3-319-25283-4_6

Proof. Suppose $\lambda, \mu > 0$ and $x, y, z \in X$.

(i) Since $|(x, x)(I_i)| = 0$ and $\varphi(0) = 0$, we get $w_\lambda(x, x) = \sup\{0\} = 0$.
Now, we prove that if $\lambda > 0$ and $w_\lambda(x, y) = 0$, then $t \mapsto d(x(t), y(t))$ is a
constant function on I. This is a consequence of (5.1.10) provided we show
that $|(x, y)(I_{s,t})| = 0$ for all $I_{s,t} = \{s, t\} \subset I$. In fact, definition (6.1.1) implies

$$\varphi\left(\frac{|(x, y)(I_{s,t})|}{\lambda}\right) \leq w_\lambda(x, y). \tag{6.1.2}$$

Since $w_\lambda(x, y) = 0$, and $\varphi(u) = 0$ iff $u = 0$, we find $|(x, y)(I_{s,t})| = 0$.
If $x, y \in \dot{X}$, then $d(x(t), y(t)) = d(x(a), y(a)) = d(x_0, x_0) = 0$ (i.e., $x(t) = y(t)$ for all $t \in I$), and so, $x = y$. This shows that w is strict on $\dot{X}$.

(ii) Clearly, $w_\lambda(x, y) = w_\lambda(y, x)$, because $|(x, y)(I_i)| = |(y, x)(I_i)|$ in (6.1.1).

(iii) , (iv) Since φ is nondecreasing on $[0, \infty)$, given $n \in \mathbb{N}$ and $\{t_i\}_0^n \prec I$,
inequalities (5.1.5) and (1.3.6) yield (cf. also (1.3.7))

$$\sum_{i=1}^n \varphi\left(\frac{|(x, y)(I_i)|}{\lambda + \mu}\right) \leq \sum_{i=1}^n \varphi\left(\frac{\lambda}{\lambda + \mu} \cdot \frac{|(x, z)(I_i)|}{\lambda} + \frac{\mu}{\lambda + \mu} \cdot \frac{|(z, y)(I_i)|}{\mu}\right)$$

$$\leq \sum_{i=1}^n \varphi\left(\frac{|(x, z)(I_i)|}{\lambda}\right) + \sum_{i=1}^n \varphi\left(\frac{|(z, y)(I_i)|}{\mu}\right)$$

$$\leq w_\lambda(x, z) + w_\mu(z, y).$$

This implies the inequality in axiom (iii). If φ is convex, the convexity property
(iv) of w is established as in (1.3.8).

To see that w is continuous from the right, it suffices to obtain the inequality
$w_\lambda(x, y) \leq w_{\lambda+0}(x, y)$. Given $n \in \mathbb{N}$, $\{t_i\}_0^n \prec I$, and $\mu > \lambda$, we have

$$\sum_{i=1}^n \varphi\left(\frac{|(x, y)(I_i)|}{\mu}\right) \leq w_\mu(x, y).$$

Passing to the limit as $\mu \to \lambda + 0$, we get $1/\mu \to (1/\lambda) - 0$, and so,

$$\sum_{i=1}^n \varphi\left(\frac{|(x, y)(I_i)|}{\lambda}\right) \leq w_{\lambda+0}(x, y)$$

by virtue of the continuity of φ. It remains to take the supremum as in (6.1.1). □

If $x^\circ(t) = x^\circ$ for all $t \in I$, where $x^\circ \in M$, then $|(x, x^\circ)(I_i)| = |x(I_i)|$ in (6.1.1),
and so, the value $w_\lambda(x, x^\circ)$ is independent of x°:

$$w_\lambda(x, x^\circ) = \sup\left\{\sum_{i=1}^n \varphi\left(\frac{d(x(t_i), x(t_{i-1}))}{\lambda}\right) : n \in \mathbb{N} \text{ and } \{t_i\}_0^n \prec I\right\}.$$

The quantity $V_\varphi(x, I) = w_1(x, x^\circ) \equiv w_1^\varphi(x, x^\circ, I)$ is known as the φ-*variation* of $x \in X$ on I in the sense of N. Wiener and L.C. Young, and the mapping x is said to be *of bounded φ-variation* provided $V_\varphi(x, I) < \infty$. Note that if we set $\varphi_\lambda(u) = \varphi(u/\lambda)$ for $u \geq 0$ and $\lambda > 0$, then $w_\lambda(x, x^\circ) = V_{\varphi_\lambda}(x, I)$.

We denote by $\mathrm{BV}_\varphi(I; M) = X_w^* \equiv X_w^*(x^\circ)$ the modular space of mappings $x : I \to M$ of bounded (generalized) φ-variation on I.

Lemma 6.1.3. $\mathrm{BV}_\varphi(I; M) \subset \mathrm{Reg}(I; M)$.

Proof. Let $x \in \mathrm{BV}_\varphi(I; M)$, so that $w_\lambda(x, x^\circ) < \infty$ for some $\lambda > 0$. First, we note that x is a bounded mapping on I: in fact, by virtue of (6.1.2),

$$|x(I)| = |(x, x^\circ)(I)| \leq \lambda \varphi_+^{-1}(w_\lambda(x, x^\circ)) < \infty, \qquad (6.1.3)$$

where φ_+^{-1} is the right inverse of φ (see Example 3.3.10). To see that x is regulated, we apply Theorem 5.1.3. To do this, we show that $x \in X_{w^\mathbb{N}}^{\mathrm{fin}}$, i.e., $w_\varepsilon^\mathbb{N}(x, x^\circ) < \infty$ for all $\varepsilon > 0$. If $\varepsilon \geq |x(I)|$, then $w_\varepsilon^\mathbb{N}(x, x^\circ) = 0$, so suppose $0 < \varepsilon < |x(I)|$. Given $n \in \mathbb{N}$ and $\{I_i\}_1^n \prec I$ such that $\min_{1 \leq i \leq n} |x(I_i)| > \varepsilon$, by the monotonicity of φ,

$$w_\lambda(x, x^\circ) \geq \sum_{i=1}^{n} \varphi(|x(I_i)|/\lambda) \geq \sum_{i=1}^{n} \varphi(\varepsilon/\lambda) = n\varphi(\varepsilon/\lambda),$$

i.e., $n \leq w_\lambda(x, x^\circ)/\varphi(\varepsilon/\lambda)$, and so,

$$w_\varepsilon^\mathbb{N}(x, x^\circ) \leq \frac{w_\lambda(x, x^\circ)}{\varphi(\varepsilon/\lambda)} < \infty, \qquad 0 < \varepsilon < |x(I)|. \qquad (6.1.4)$$

$\square$

Remark 6.1.4. For a convex function φ, w from (6.1.1) is convex, and so, $X_w^0 = X_w^*$ (see Sect. 2.1 and Proposition 1.2.3(c)). If φ is nonconvex, the problem of characterization of the spaces X_w^0 and X_w^{fin} is still unsolved even in the case when $M = \mathbb{R}$ and, in particular, it is not known whether the space(s) X_w^0 and/or X_w^{fin} coincide(s) with $X_w^* = \mathrm{BV}_\varphi(I; M)$ or not.

Lemma 6.1.5. *Given $x, y \in \mathrm{BV}_\varphi(I; M)$, we have:*

(a) $d^1(x, y) = d(x(a), y(a)) + d_w^0(x, y)$ *is a metric on* $\mathrm{BV}_\varphi(I; M)$, *and*

$$d_\infty(x, y) \leq d(x(a), y(a)) + d_w^0(x, y) \cdot \varphi_+^{-1}(d_w^0(x, y));$$

(b) *if, in addition, φ is convex, then* $d_M(x, y) = d(x(a), y(a)) + d_w^*(x, y)$ *is a metric on* $\mathrm{BV}_\varphi(I; M)$ *such that* $d_\infty(x, y) \leq \max\{1, \varphi^{-1}(1)\} d_M(x, y)$.

Proof. (a) By Lemma 6.1.2 and Theorem 2.2.1, d_w^0 is a pseudometric on the space $X_w^* = \mathrm{BV}_\varphi(I; M)$, and $d_w^0(x, y) < \infty$ for $x, y \in X_w^*$. For every $\lambda > d_w^0(x, y)$, we have $w_\lambda(x, y) \leq \lambda$, which together with (6.1.2) implies $|(x, y)(I_{s,t})| \leq \lambda \varphi_+^{-1}(\lambda)$, and so, passing to the limit as $\lambda \to d_w^0(x, y)$, we get $|(x, y)(I_{s,t})| \leq d_w^0(x, y) \cdot \varphi_+^{-1}(d_w^0(x, y))$ for all $I_{s,t} = \{s, t\} \subset I$. Now, the inequality in (a) follows from (5.1.10) with $s = a$.

(b) is established similarly: applying Theorem 2.3.1 and noting that $\lambda > d_w^*(x, y)$ implies $w_\lambda(x, y) \leq 1$, and φ is strictly increasing on $[0, \infty)$ (see Appendix A.1), we get $|(x, y)(I_{s,t})| \leq \varphi^{-1}(1)\lambda$, and so, $|(x, y)(I_{s,t})| \leq \varphi^{-1}(1)d_w^*(x, y)$ for all $s, t \in I$. It follows from (5.1.10) that

$$d(x(t), y(t)) \leq d(x(a), y(a)) + \varphi^{-1}(1)d_w^*(x, y) \quad \text{for all } t \in I. \tag{6.1.5}$$

$\square$

Theorem 6.1.6. (a) *Given $x, y \in X$, $a \leq s < t \leq b$, and $\lambda > 0$, we have*

$$w_\lambda(x, y, [a, s]) + w_\lambda(x, y, [s, t]) \leq w_\lambda(x, y, [a, t]).$$

(b) *If $\{x_n\}$, $\{y_n\} \subset X$, $x, y \in X$, $x_n \to x$ and $y_n \to y$ pointwise on I, then $w_\lambda(x, y) \leq \liminf_{n \to \infty} w_\lambda(x_n, y_n)$ for all $\lambda > 0$, and $d_w^0(x, y) \leq \liminf_{n \to \infty} d_w^0(x_n, y_n)$. Furthermore, if φ is convex, then $d_w^*(x, y) \leq \liminf_{n \to \infty} d_w^*(x_n, y_n)$.*

(c) *If the metric space (M, d) is complete, then $(\mathrm{BV}_\varphi(I; M), d^1)$ is a complete metric space. A similar assertion holds for $(\mathrm{BV}_\varphi(I; M), \mathrm{d}_M)$ if φ is convex.*

Proof. (a) For $n, m \in \mathbb{N}$ and partitions $\{t_i\}_0^n \prec [a, s]$ and $\{t_j'\}_0^m \prec [s, t]$, $\{t_i\}_0^n \cup \{t_j'\}_0^m$ is a partition of $[a, t]$, and so, if $I_i = \{t_{i-1}, t_i\}$ and $I_j' = \{t_{j-1}', t_j'\}$, then

$$\sum_{i=1}^n \varphi\left(\frac{|(x, y)(I_i)|}{\lambda}\right) + \sum_{j=1}^m \varphi\left(\frac{|(x, y)(I_j')|}{\lambda}\right) \leq w_\lambda(x, y, [a, t]).$$

It remains to take the corresponding suprema in the left-hand side.

(b) 1. First, we note that, given $n \in \mathbb{N}$ and $I_{s,t} = \{s, t\} \subset I$, we have

$$\left||(x_n, y_n)(I_{s,t})| - |(x, y)(I_{s,t})|\right| \leq d(x_n(t), x(t)) + d(x_n(s), x(s))$$
$$+ d(y_n(t), y(t)) + d(y_n(s), y(s)). \tag{6.1.6}$$

In fact, by virtue of (5.1.5) and (5.1.8),

$$|(x_n, y_n)(I_{s,t})| \leq |(x_n, x)(I_{s,t})| + |(x, y)(I_{s,t})| + |(y, y_n)(I_{s,t})|$$
$$\leq d(x_n(t), x(t)) + d(x_n(s), x(s)) + |(x, y)(I_{s,t})|$$
$$+ d(y(t), y_n(t)) + d(y(s), y_n(s)).$$

Inequality (6.1.6) follows if we exchange x_n and x, and y_n and y.

From the pointwise convergence of x_n to x, y_n to y, and (6.1.6), we find

$$|(x_n, y_n)(I_{s,t})| \to |(x, y)(I_{s,t})| \quad \text{as } n \to \infty \text{ for all } s, t \in I.$$

Definition (6.1.1) implies that if $m \in \mathbb{N}$ and $\{t_i\}_0^m \prec I$, then

$$\sum_{i=1}^{m} \varphi\left(\frac{|(x_n, y_n)(I_i)|}{\lambda}\right) \le w_\lambda(x_n, y_n) \quad \text{for all } n \in \mathbb{N}.$$

Passing to the limit inferior as $n \to \infty$, by the continuity of φ, we get

$$\sum_{i=1}^{m} \varphi\left(\frac{|(x, y)(I_i)|}{\lambda}\right) \le \liminf_{n \to \infty} w_\lambda(x_n, y_n).$$

Since $m \in \mathbb{N}$ and $\{t_i\}_0^m \prec I$ are arbitrary, we are through.

2. In order to prove the second inequality, it suffices to assume that the quantity $\lambda \equiv \liminf_{n \to \infty} d_w^0(x_n, y_n)$ is finite. It follows that $d_w^0(x_{n_k}, y_{n_k}) \to \lambda$ as $k \to \infty$ for some subsequence $\{n_k\}_{k=1}^{\infty}$ of $\{n\}_{n=1}^{\infty}$. So, given $\varepsilon > 0$, there is $k_0 = k_0(\varepsilon) \in \mathbb{N}$ such that $d_w^0(x_{n_k}, y_{n_k}) < \lambda + \varepsilon$ for all $k \ge k_0$. By the definition of d_w^0, $w_{\lambda+\varepsilon}(x_{n_k}, y_{n_k}) \le \lambda + \varepsilon$ for $k \ge k_0$. Since $x_{n_k} \to x$ and $y_{n_k} \to y$ pointwise on I, we find

$$w_{\lambda+\varepsilon}(x, y) \le \liminf_{k \to \infty} w_{\lambda+\varepsilon}(x_{n_k}, y_{n_k}) \le \lambda + \varepsilon.$$

This means that $d_w^0(x, y) \le \lambda + \varepsilon$ for all $\varepsilon > 0$, and so, $d_w^0(x, y) \le \lambda$.

Similar arguments apply to prove the third inequality in (b).

(c) Let $\{x_n\}$ be a Cauchy sequence in $BV_\varphi(I; M)$, i.e.,

$$d^1(x_n, x_m) = d(x_n(a), x_m(a)) + d_w^0(x_n, x_m) \to 0 \quad \text{as } n, m \to \infty. \quad (6.1.7)$$

The estimate in Lemma 6.1.5(a) implies that, for every $t \in I$, $\{x_n(t)\}$ is a Cauchy sequence in M. By the completeness of M, there exists a mapping $x : I \to M$ such that $d(x_n(t), x(t)) \to 0$ as $n \to \infty$ for all $t \in I$. Since $x_n \to x_n$ and $x_m \to x$ pointwise on I as $m \to \infty$, item (b) of this Theorem yields

$$d_w^0(x_n, x) \le \liminf_{m \to \infty} d_w^0(x_n, x_m) \le \lim_{m \to \infty} d^1(x_n, x_m) < \infty \quad \text{for all } n \in \mathbb{N}.$$

It follows from (6.1.7) that

$$\limsup_{n \to \infty} d_w^0(x_n, x) \le \lim_{n \to \infty} \lim_{m \to \infty} d^1(x_n, x_m) = 0,$$

and so, $d^1(x_n, x) \to 0$ as $n \to \infty$. In order to show that $x \in \mathrm{BV}_\varphi(I; M)$, we note that, by the triangle inequality for d_w^0,

$$|d_w^0(x_n, x^\circ) - d_w^0(x_m, x^\circ)| \le d_w^0(x_n, x_m) \to 0 \quad \text{as} \quad n, m \to \infty.$$

Hence $\{d_w^0(x_n, x^\circ)\}$ is a Cauchy sequence in $\mathbb{R}$, and so, it is convergent (and bounded). Since x_n converges to x pointwise on I, we get

$$d_w^0(x, x^\circ) \le \liminf_{n\to\infty} d_w^0(x_n, x^\circ) = \lim_{n\to\infty} d_w^0(x_n, x^\circ) < \infty.$$

By Theorem 2.2.1, $x \sim x^\circ$, i.e., $x \in X_w^*(x^\circ) = \mathrm{BV}_\varphi(I; M)$.
The last assertion in (c) is proved similarly. $\square$

A generalization of the classical Helly selection principle for real functions of bounded Jordan variation is given in the following

Corollary 6.1.7. *A pointwise precompact sequence of mappings $\{x_n\} \subset \mathrm{BV}_\varphi(I; M)$, satisfying $C_\lambda \equiv \sup_{n\in\mathbb{N}} V_{\varphi_\lambda}(x_n, I) < \infty$ for some $\lambda > 0$, contains a pointwise convergent subsequence whose pointwise limit x belongs to the space $\mathrm{BV}_\varphi(I; M)$.*

Proof. In order to apply Theorem 5.2.1 with $y_n \equiv x^\circ$, it suffices to verify condition (5.2.1). Inequality (6.1.3) implies

$$|x_n(I)| \le \lambda\varphi_+^{-1}(w_\lambda(x_n, x^\circ)) = \lambda\varphi_+^{-1}(V_{\varphi_\lambda}(x_n, I)) \le \lambda\varphi_+^{-1}(C_\lambda), \quad n \in \mathbb{N}.$$

For $\varepsilon \ge |x_n(I)|$, we have $w_\varepsilon^{\mathbb{N}}(x_n, x^\circ) = 0$, and if $0 < \varepsilon < |x_n(I)|$, then, by virtue of (6.1.4),

$$w_\varepsilon^{\mathbb{N}}(x_n, x^\circ) \le \frac{w_\lambda(x_n, x^\circ)}{\varphi(\varepsilon/\lambda)} = \frac{V_{\varphi_\lambda}(x_n, I)}{\varphi(\varepsilon/\lambda)} \le \frac{C_\lambda}{\varphi(\varepsilon/\lambda)}.$$

It follows that

$$\sup_{n\in\mathbb{N}} w_\varepsilon^{\mathbb{N}}(x_n, x^\circ) \le \frac{C_\lambda}{\varphi(\varepsilon/\lambda)} < \infty \quad \text{for all} \quad \varepsilon > 0,$$

and so, condition (5.2.1) is satisfied. By Theorem 5.2.1, a subsequence $\{x_{n_k}\}$ of $\{x_n\}$ converges pointwise on I to a mapping $x \in \mathrm{Reg}(I; M)$. Applying Theorem 6.1.6(b), we get

$$w_\lambda(x, x^\circ) \le \liminf_{k\to\infty} w_\lambda(x_{n_k}, x^\circ) = \liminf_{k\to\infty} V_{\varphi_\lambda}(x_{n_k}, I) \le C_\lambda < \infty,$$

which implies $x \in X_w^*(x^\circ) = \mathrm{BV}_\varphi(I; M)$. $\square$

6.2 Lipschitzian Operators

Let (M, d) and (N, d) be two metric spaces (with different metrics d, in general). Denote by $\mathbb{X} = M^N$ the set of all operators $T : N \to M$ mapping N into M. Given $\lambda > 0$ and $T, S \in \mathbb{X}$, we set $\mathbb{W}_\lambda(T, S) = \mathbb{D}(T, S)/\lambda$, where

$$\mathbb{D}(T, S) = \sup \left\{ \frac{|(T, S)(I_{x,y})|}{d(x, y)} : I_{x,y} = \{x, y\} \subset N,\ x \neq y \right\}$$

and (following the pattern of (5.1.1))

$$|(T, S)(I_{x,y})| = \sup_{v \in M} \left| d(Tx, v) + d(Sy, v) - d(Ty, v) - d(Sx, v) \right|.$$

If, in addition, $(M, d, +)$ is a metric semigroup, we may replace the last quantity (as in (5.1.2)) by

$$|(T, S)(I_{x,y})| = d(Tx + Sy, Ty + Sx), \qquad x, y \in N. \tag{6.2.1}$$

For every $\lambda > 0$, $\mathbb{W}_\lambda$ is an extended pseudometric on $\mathbb{X}$, and so (see Sect. 1.3.2), $\mathbb{W}$ is a convex pseudomodular on $\mathbb{X}$, and a convex modular on $\mathring{\mathbb{X}} = \{T : N \to M \mid Tu_0 = v_0\}$, where elements $u_0 \in N$ and $v_0 \in M$ are fixed.

If $C : N \to M$ is a constant operator (i.e., $Cx = Cy$ for all $x, y \in N$), then we have $|(T, C)(I_{x,y})| = d(Tx, Ty)$, $x, y \in N$, and so, $\mathbb{W}_\lambda(T, C) = \mathbb{L}(T)/\lambda$, where

$$\mathbb{L}(T) = \mathbb{D}(T, C) = \sup \{ d(Tx, Ty)/d(x, y) : x, y \in N, x \neq y \} \tag{6.2.2}$$

is the *least Lipschitz constant* of $T \in \mathbb{X}$. Clearly, $\mathbb{L}(T)$ is independent of C. It follows that the modular space $\mathbb{X}_\mathbb{W}^* = \mathbb{X}_\mathbb{W}^*(C) = \{T \in \mathbb{X} : \mathbb{L}(T) < \infty\}$ is the set of all *Lipschitzian operators* from N into M, which is denoted by $\mathrm{Lip}(N; M)$. The function $d_\mathbb{W}^*(T, S) = \mathbb{D}(T, S)$ from (2.3.3) is a pseudometric on $\mathrm{Lip}(N; M)$, and a metric on $\mathrm{Lip}_0(N; M) = \mathring{\mathbb{X}}_\mathbb{W}^* = \{T \in \mathrm{Lip}(N; M) : Tu_0 = v_0\}$.

Given $T, S \in \mathrm{Lip}(N; M)$, the following inequalities hold:

$$|\mathbb{L}(T) - \mathbb{L}(S)| \leq \mathbb{D}(T, S) \leq \mathbb{L}(T) + \mathbb{L}(S),$$

and, by virtue of (5.1.10), for every $x, y \in N$,

$$|d(Tx, Sx) - d(Ty, Sy)| \leq |(T, S)(I_{x,y})| \leq \mathbb{D}(T, S) d(x, y). \tag{6.2.3}$$

6.3   Superposition Operators

Suppose $(N, d, +)$ and $(M, d, +)$ are two metric semigroups with (generally, different) zeros 0. Let $T : N \to M$ be an *additive operator*, i.e., it satisfies the *Cauchy equation* $T(x + y) = Tx + Ty$ for all $x, y \in N$. Clearly, we have $T(0) = 0$: in fact, $d(0, T(0)) = d(T(0), T(0) + T(0)) = d(T(0), T(0)) = 0$. The *zero* operator $\mathbf{0} : N \to M$ is the constant operator such that $\mathbf{0}x = 0 \in M$ for all $x \in N$. We denote by $LA(N; M)$ the set of all *additive* operators from $\mathrm{Lip}(N; M)$.

Let φ be a convex φ-function. Since $\mathbb{D}$ (based on (6.2.1)) is a metric on $LA(N; M)$, we may replace M by $LA(N; M)$, and d—by $\mathbb{D}$ in Lemma 6.1.5(b), and consider the space $BV_\varphi(I; LA(N; M))$ equipped with the metric

$$\mathbb{D}_{N,M}(T, S) = \mathbb{D}(T(a), S(a)) + \mathbb{D}_w^*(T, S), \qquad T, S \in BV_\varphi(I; LA(N; M)), \qquad (6.3.1)$$

where $\mathbb{D}_w^*(T, S) = \inf\{\lambda > 0 : w_\lambda^{\mathbb{D}}(T, S) \leq 1\}$ (see (2.3.3)) and

$$w_\lambda^{\mathbb{D}}(T, S) = \sup\left\{ \sum_{i=1}^n \varphi\left(\frac{\mathbb{D}(T(t_i) + S(t_{i-1}), T(t_{i-1}) + S(t_i))}{\lambda} \right) : n \in \mathbb{N} \text{ and } \{t_i\}_0^n \prec I \right\}.$$

Also, we denote by $\mathbf{0}$ the mapping $\mathbf{0}(t) = \mathbf{0}$, $t \in I$, so that $\mathbf{0} \in BV_\varphi(I; LA(N; M))$.

Recall that, given a mapping $h : I \times N \to M$, the operator $H \equiv H_h : N^I \to M^I$, defined by $(Hx)(t) = h(t, x(t))$ for all $t \in I$ and $x : I \to N$, is said to be the *superposition (Nemytskii) operator* with the *generator* h.

In the next theorem, we make use of the assumptions above concerning N, M, and φ. It is to be noted that, in Theorem 6.3.1, in order to construct metric spaces $(BV_\varphi(I; N), d_N)$ and $(BV_\varphi(I; M), d_M)$, we apply (5.1.2) instead of (5.1.1).

Theorem 6.3.1. *Given $T \in BV_\varphi(I; LA(N; M))$, define the mapping $h : I \times N \to M$ by the rule: $h(t, x) = T(t)x$ for all $t \in I$ and $x \in N$. Then the superposition operator H, generated by h, maps the metric space $(BV_\varphi(I; N), d_N)$ into the metric space $(BV_\varphi(I; M), d_M)$ and is Lipschitzian, and the following inequality holds:*

$$d_M(Hx, Hy) \leq \gamma(\varphi)\mathbb{D}_{N,M}(T, \mathbf{0}) d_N(x, y) \quad \text{for all } x, y \in BV_\varphi(I; N), \qquad (6.3.2)$$

where $\gamma(\varphi) = \max\{1, 2\varphi^{-1}(1)\}$.

Proof. Taking into account Lemma 6.1.5(b), we have to estimate the quantity

$$d_M(Hx, Hy) = d((Hx)(a), (Hy)(a)) + d_w^*(Hx, Hy).$$

Since $T(a) \in \mathrm{Lip}(N; M)$, the first term is estimated from (6.2.2):

$$d((Hx)(a), (Hy)(a)) = d(T(a)x(a), T(a)y(a)) \leq \mathbb{D}(T(a), \mathbf{0})d(x(a), y(a)) = A_0 B_0,$$

where $A_0 = \mathbb{D}(T(a), \mathbf{0})$ and $B_0 = d(x(a), y(a))$. Now, it suffices to prove that

$$d_w^*(Hx, Hy) \le \lambda A + \mu B \qquad (6.3.3)$$

with $\lambda = d_w^*(x, y)$, $\mu = \mathbb{D}_w^*(T, \mathbf{0})$, $A = \sup_{t \in I} \mathbb{D}(T(t), \mathbf{0})$, and $B = d_\infty(x, y)$. In fact, noting that, in view of Lemma 6.1.5(b) and equality (6.3.1), $d_N(x, y) = B_0 + \lambda$ and $\mathbb{D}_{N,M}(T, \mathbf{0}) = A_0 + \mu$, and that, by virtue of (6.1.5),

$$B = \sup_{t \in I} d(x(t), y(t)) \le d(x(a), y(a)) + \varphi^{-1}(1)d_w^*(x, y) = B_0 + \varphi^{-1}(1)\lambda,$$

and similarly (replacing d by $\mathbb{D}$ and x—by T in (6.1.5)),

$$A = \sup_{t \in I} \mathbb{D}(T(t), \mathbf{0}) \le \mathbb{D}(T(a), \mathbf{0}) + \varphi^{-1}(1)\mathbb{D}_w^*(T, \mathbf{0}) = A_0 + \varphi^{-1}(1)\mu,$$

we find $d_M(Hx, Hy) \le A_0 B_0 + \lambda A + \mu B$, which is less than or equal to

$$A_0 B_0 + \lambda(A_0 + \varphi^{-1}(1)\mu) + \mu(B_0 + \varphi^{-1}(1)\lambda) \le \gamma(\varphi)(A_0 + \mu)(B_0 + \lambda).$$

The last expression is exactly the right-hand side of the inequality (6.3.2).

Now we establish inequality (6.3.3). Given $I_{s,t} = \{s, t\} \subset I$, the additivity property of the operator $T(t)$ implies the equality

$$[T(t)x(t) + T(s)y(s)] + [T(t)(x(s) + y(t))] + [T(t)y(s) + T(s)x(s)]$$
$$= [T(s)x(s) + T(t)y(t)] + [T(t)(x(t) + y(s))] + [T(t)x(s) + T(s)y(s)].$$

$$(6.3.4)$$

For $k = 0, 1, 2$, denote by ℓ_k (by r_k) the $(k+1)$-th term in square brackets in the left-hand (right-hand) side of this equality, so that $\ell_0 + \ell_1 + \ell_2 = r_0 + r_1 + r_2$. Applying inequality (A.3.1) from Appendix A.3, we have

$$d(\ell_0, r_0) = d(\ell_0 + \ell_1 + \ell_2, r_0 + \ell_1 + \ell_2) = d(r_0 + r_1 + r_2, r_0 + \ell_1 + \ell_2)$$
$$= d(r_1 + r_2, \ell_1 + \ell_2) \le d(r_1, \ell_1) + d(r_2, \ell_2). \qquad (6.3.5)$$

Since $(Hx)(t) = T(t)x(t)$ for all $t \in I$ and $x \in N^I$, (5.1.2) implies

$$|(Hx, Hy)(I_{s,t})| = d\big((Hx)(t) + (Hy)(s), (Hx)(s) + (Hy)(t)\big)$$
$$= d\big(T(t)x(t) + T(s)y(s), T(s)x(s) + T(t)y(t)\big).$$

By (6.3.4), (6.3.5), (6.2.2), and (6.2.1), this is less than or equal to

$$d\big(T(t)(x(t) + y(s)), T(t)(x(s) + y(t))\big)$$
$$+ d\big(T(t)x(s) + T(s)y(s), T(t)y(s) + T(s)x(s)\big)$$
$$\leq \mathbb{D}(T(t), \mathbf{0})d\big(x(t) + y(s), x(s) + y(t)\big) + |(T(t), T(s))(I_{x(s), y(s)})|.$$

Now, it follows from (5.1.2) and (6.2.3) that

$$|(Hx, Hy)(I_{s,t})| \leq \mathbb{D}(T(t), \mathbf{0})|(x, y)(I_{s,t})| + \mathbb{D}(T(t), T(s))d(x(s), y(s))$$
$$\leq A \cdot |(x, y)(I_{s,t})| + |(T, \mathbf{0})(I_{s,t})| \cdot B, \qquad (6.3.6)$$

where, by virtue of (5.1.2),

$$|(T, \mathbf{0})(I_{s,t})| = \mathbb{D}(T(t) + \mathbf{0}(s), T(s) + \mathbf{0}(t)) = \mathbb{D}(T(t), T(s)).$$

If $AB = 0$, then $A = 0$ or $B = 0$ and, since $\mathbb{D}$ is a metric on $\mathrm{LA}(N; M)$ and d is a metric on N, we have $T(t) = \mathbf{0}$ for all $t \in I$, or $x(t) = y(t)$ for all $t \in I$, and so, the left- and right-hand sides in (6.3.3) are equal to zero.

Suppose $AB \neq 0$. Let $\lambda\mu \neq 0$. From (6.3.6), the convexity of φ, and (6.1.1), given a partition $\{t_i\}_0^n \prec I$ (so that $I_i = \{t_{i-1}, t_i\}$), we find

$$\sum_{i=1}^{n} \varphi\left(\frac{|(Hx, Hy)(I_i)|}{\lambda A + \mu B}\right) \leq \frac{\lambda A}{\lambda A + \mu B} \sum_{i=1}^{n} \varphi\left(\frac{|(x, y)(I_i)|}{\lambda}\right)$$
$$+ \frac{\mu B}{\lambda A + \mu B} \sum_{i=1}^{n} \varphi\left(\frac{|(T, \mathbf{0})(I_i)|}{\mu}\right)$$
$$\leq \frac{\lambda A}{\lambda A + \mu B} w_\lambda(x, y) + \frac{\mu B}{\lambda A + \mu B} w_\mu^{\mathbb{D}}(T, \mathbf{0}),$$

and so, again by (6.1.1),

$$w_{\lambda A + \mu B}(Hx, Hy) \leq \frac{\lambda A}{\lambda A + \mu B} w_\lambda(x, y) + \frac{\mu B}{\lambda A + \mu B} w_\mu^{\mathbb{D}}(T, \mathbf{0}). \qquad (6.3.7)$$

By Lemma 6.1.2, functions $\xi \mapsto w_\xi(x, y)$ and $\eta \mapsto w_\eta^{\mathbb{D}}(T, \mathbf{0})$ are continuous from the right on $(0, \infty)$, and so, applying Theorem 2.3.2(c), we get $w_\lambda(x, y) \leq 1$ and $w_\mu^{\mathbb{D}}(T, \mathbf{0}) \leq 1$. Inequality (6.3.7) implies $w_{\lambda A + \mu B}(Hx, Hy) \leq 1$, and now inequality (6.3.3) follows from definition (2.3.3).

Assume that $\lambda = 0$. Then $|(x, y)(I_{s,t})| \leq \varphi^{-1}(1)d_w^*(x, y) = 0$ (see the proof of Lemma 6.1.5(b)), and so, (6.3.6) implies

$$|(Hx, Hy)(I_{s,t})| \leq |(T, \mathbf{0})(I_{s,t})| \cdot B \quad \text{for all } s, t \in I. \qquad (6.3.8)$$

If $\mu \neq 0$, then it follows from (6.3.8) that $w_{\mu B}(Hx, Hy) \leq w_\mu^{\mathbb{D}}(T, 0) \leq 1$ (in place of (6.3.7)), and so, $d_w^*(Hx, Hy) \leq \mu B$ as asserted in (6.3.3). If $\mu = 0$, then, similar to the above, $|(T, 0)(I_{s,t})| \leq \varphi^{-1}(1)\mathbb{D}_w^*(T, 0) = 0$, which implies that the right-hand side in (6.3.8) is equal to zero for all $s, t \in I$, and so, $w_\eta(Hx, Hy) = 0$ for all $\eta > 0$, i.e., $d_w^*(Hx, Hy) = 0$ in (6.3.3). In the final case when $\lambda \neq 0$ and $\mu = 0$, we find $|(T, 0)(I_{s,t})| = 0$ in (6.3.6), which yields $w_{\lambda A}(Hx, Hy) \leq w_\lambda(x, y) \leq 1$. This gives $d_w^*(Hx, Hy) \leq \lambda A$, establishing (6.3.3).

In order to see that H maps $\mathrm{BV}_\varphi(I; N)$ into $\mathrm{BV}_\varphi(I; M)$, we set $y(t) = 0$ for all $t \in I$ in (6.3.2): since $Hy = 0$, we find

$$d_w^*(Hx, 0) \leq \mathsf{d}_M(H(x), H(0)) \leq \gamma(\varphi)\mathbb{D}_{N,M}(T, 0)\mathsf{d}_N(x, 0) < \infty.$$

By Theorem 2.3.1, $Hx \sim 0 \sim x^\circ$, and so, $Hx \in X_w^* = \mathrm{BV}_\varphi(I; M)$.   □

Remark 6.3.2. If, in Theorem 6.3.1, $N = M$ is a complete metric semigroup and $T \in \mathrm{BV}_\varphi(I; \mathrm{LA}(M; M))$ is such that $\gamma(\varphi)\mathbb{D}_{M,M}(T, 0) < 1$, then Theorem 6.1.6(c) implies that $(\mathrm{BV}_\varphi(I; M), \mathsf{d}_M)$ is also a complete metric semigroup, and so, by the Banach Contraction Theorem, there is a unique $x \in \mathrm{BV}_\varphi(I; M)$ such that $x = Hx$, i.e., x is the unique solution of the functional equation $x(t) = T(t)x(t)$ in M ($t \in I$).

6.4   Selections of Bounded Variation

Given a metric space (M, d), let us consider a particular case of the pseudomodular $w_\lambda^\varphi(x, y)$ from Definition 6.1.1 on the set $X = M^I$ when $\varphi(u) = \mathrm{id}(u) = u$, i.e., $w_\lambda^{\mathrm{id}}(x, y, I) = (1/\lambda)\,w(x, y, I)$, where $\lambda > 0$, $x, y \in X$, and

$$w(x, y, I) = \sup\left\{\sum_{i=1}^n |(x, y)(I_i)| : n \in \mathbb{N} \text{ and } \{t_i\}_0^n \prec I\right\}. \tag{6.4.1}$$

Clearly (cf. the proof of Lemma 5.1.2), w is a pseudometric on X.

Theorem 6.1.6(a) can be refined as follows.

Lemma 6.4.1. *Given $x, y \in X$ and $a \leq s < t \leq b$, we have:*

$$w(x, y, [a, s]) + w(x, y, [s, t]) = w(x, y, [a, t]) \quad \textit{(additivity of w)}.$$

Proof. In order to verify the inequality $\geq$, let $\{t_i\}_0^n \prec [a, t]$ be a partition of the interval $[a, t]$, so that $I_i = \{t_{i-1}, t_i\}$, $i = 1, \ldots, n$, and $t_{k-1} \leq s \leq t_k$ for some $1 \leq k \leq n$. By virtue of (5.1.1) (or (5.1.2), or (5.1.3)),

$$|(x, y)(I_k)| = |(x, y)(I_{t_{k-1}, t_k})| \leq |(x, y)(I_{t_{k-1}, s})| + |(x, y)(I_{s, t_k})|, \tag{6.4.2}$$

whence

$$\sum_{i=1}^{n} |(x,y)(I_i)| \le \sum_{i=1}^{k-1} |(x,y)(I_i)| + |(x,y)(I_k)| + \sum_{i=k+1}^{n} |(x,y)(I_i)|$$

$$\le \mathrm{w}(x,y,[a,s]) + \mathrm{w}(x,y,[s,t]),$$

and it remains to take the supremum over all $n \in \mathbb{N}$ and $\{t_i\}_0^n \prec [a,t]$.   □

The quantity $V(x,I) = V_{\mathrm{id}}(x,I) = \mathrm{w}(x,x^\circ,I)$, i.e.,

$$V(x,I) = \sup \left\{ \sum_{i=1}^{n} d(x(t_i), x(t_{i-1})) : n \in \mathbb{N} \text{ and } \{t_i\}_0^n \prec I \right\},$$

is the usual (C. Jordan) *variation* of mapping $x \in X$ on I, and x is said to be *of bounded variation* on I provided $V(x,I) < \infty$.

We denote by $\mathrm{BV}(I;M) = \mathrm{BV}_{\mathrm{id}}(I;M)$ the modular space of all mappings $x : I \to M$ of bounded variation on I. The set of discontinuity points of an $x \in \mathrm{BV}(I;M)$ is an at most countable subset of I: in fact, given $a \le s \le t \le b$, by virtue of Lemma 6.4.1, we have

$$d(x(t),x(s)) \le V(x,[s,t]) = V(x,[a,t]) - V(x,[a,s]), \tag{6.4.3}$$

where $t \mapsto V(x,[a,t])$ is a nondecreasing function on I, and hence regulated.

Remark 6.4.2. If φ is a convex φ-function on $[0,\infty)$, then

$$\mathrm{BV}(I;M) \subset \mathrm{BV}_\varphi(I;M). \tag{6.4.4}$$

In fact, since φ is superadditive (Appendix A.1), $V_\varphi(x,I) \le \varphi(V(x,I))$ for all mappings $x \in \mathrm{BV}(I;M)$. Moreover, if $\varphi'(0) > 0$, the two spaces in (6.4.4) coincide: given $x \in \mathrm{BV}_\varphi(I;M)$, there is $\lambda > 0$ such that $V_{\varphi_\lambda}(x,I) = \mathrm{w}_\lambda^\varphi(x,x^\circ) < \infty$ and, since $\varphi'(0) = \inf_{u>0} \varphi(u)/u$, we have $u \le \lambda \varphi_\lambda(u)/\varphi'(0)$ for all $u \ge 0$, which implies $V(x,I) \le \lambda V_{\varphi_\lambda}(x,I)/\varphi'(0)$, and so, $x \in \mathrm{BV}(I;M)$.

If φ is not necessarily convex and $\limsup_{u\to+0} \varphi(u)/u < \infty$, then inclusion (6.4.4) holds, and the reverse inclusion in (6.4.4) holds if $\limsup_{u\to+0} u/\varphi(u) < \infty$.

Recall that a mapping $F : I \to \mathscr{P}(M)$ is said to be a *multifunction*, and a mapping $f : I \to M$ is called a *selection* of F if $f(t) \in F(t)$ for all $t \in I$. By the Axiom of Choice, selections always exist. However, in what follows we are interested in selections preserving certain regularity properties of the original multifunction such as 'to be of bounded variation'.

Denote by $\mathrm{c}(M)$ the family of all nonempty compact subsets of M, equipped with the Hausdorff metric $D = D_d$ induced by metric d on M (Appendix A.2).

Theorem 6.4.3 (existence of BV selections). *Given $F \in BV(I; c(M))$, $t_0 \in I$, and $x_0 \in F(t_0)$, there exists $x \in BV(I; M)$ such that $x(t) \in F(t)$ for all $t \in I$, $x(t_0) = x_0$, and $V(x, I) \le V(F, I)$.*

Proof. Since the multifunction $F : I \to c(M)$ is of bounded variation (with respect to D), the set $E \subset I$ of points of discontinuity of F is at most countable. Denote by S the set $\{s_0, s_1, s_2\} \cup (I \cap \mathbb{Q}) \cup E$, where $s_0 = a$, $s_1 = t_0$, and $s_2 = b$. Then $S = \{s_j\}_{j=0}^{\infty}$ is a countable dense subset of I such that, given $n \ge 2$, the collection $\{s_j\}_{j=0}^{n}$ is a partition of I. Ordering the points in $\{s_j\}_{j=0}^{n}$ in ascending order and denoting the resulting collection by $\{t_i^n\}_{i=0}^{n}$, we have: $\{t_i^n\}_{i=0}^{n} \prec I$ (i.e., $a = t_0^n < t_1^n < \cdots < t_{n-1}^n < t_n^n = b$), $t_0 = t_{k_0(n)}^n$ for some $k_0(n) \in \{0, 1, \dots, n\}$, and

$$\text{for every } s \in S \text{ there is } n_0(s) \in \mathbb{N} \text{ such that } s \in \{t_i^n\}_{i=0}^{n} \text{ for all } n \ge n_0(s). \tag{6.4.5}$$

Taking into account that $F(t) \in c(M)$, $t \in I$, we define $x_i^n \in F(t_i^n)$ for all $i = 0, 1, \dots, n$ inductively as follows. First, assume that $a < t_0 < b$, and so, $k_0(n) \in \{1, \dots, n-1\}$.

(a) Set $x_{k_0(n)}^n = x_0$.
(b) If $i \in \{k_0(n) + 1, \dots, n\}$ and $x_{i-1}^n \in F(t_{i-1}^n)$ is already chosen, pick an element $x_i^n \in F(t_i^n)$ such that $d(x_{i-1}^n, x_i^n) = \text{dist}(x_{i-1}^n, F(t_i^n))$ (see (A.2.1)).
(c) If $i \in \{1, \dots, k_0(n)\}$ and $x_i^n \in F(t_i^n)$ is already chosen, pick $x_{i-1}^n \in F(t_{i-1}^n)$ such that $d(x_i^n, x_{i-1}^n) = \text{dist}(x_i^n, F(t_{i-1}^n))$.

Now, if $t_0 = a$ (then $k_0(n) = 0$), we define $x_i^n \in F(t_i^n)$ following (a) and (b), and if $t_0 = b$ (then $k_0(n) = n$), we define $x_i^n \in F(t_i^n)$ as in (a) and (c).

Let $x_n : I \to M$ ($n \ge 2$) be a step mapping of the form:

$$x_n(t) = x_{i-1}^n \quad \text{if} \quad t_{i-1}^n \le t < t_i^n, \quad i = 1, \dots, n, \quad \text{and} \quad x_n(b) = x_n^n. \tag{6.4.6}$$

We have $x_n(t_0) = x_n(t_{k_0(n)}^n) = x_{k_0(n)}^n = x_0$ and, by the additivity in Lemma 6.4.1 of $V(\cdot, J) = \text{w}(\cdot, x^\circ, J)$ in the variable interval $J \subset I$, (b) and (c) above, and the definition of the Hausdorff metric D (Appendix A.2),

$$V(x_n, I) = \sum_{i=1}^{n} V(x_n, [t_{i-1}^n, t_i^n]) = \sum_{i=1}^{n} d(x_i^n, x_{i-1}^n)$$

$$\le \sum_{i=1}^{n} D(F(t_i^n), F(t_{i-1}^n)) \le V(F, I) \quad \text{for all } n \ge 2. \tag{6.4.7}$$

It follows that $\sup_{n \ge 2} V(x_n, I) < \infty$, and so, in order to apply Corollary 6.1.7, we ought to verify that the sequence $\{x_n\}$ is pointwise precompact on I.

By (6.4.5), if $s \in S$, then $s \in \{t_i^n\}_{i=0}^{n}$ for all $n \ge n_0(s)$, and so, (a), (b) and (c) above, and definition (6.4.6) imply

$$x_n(s) \in F(s) \quad \text{for all } n \ge n_0(s). \tag{6.4.8}$$

Since $F(s) \subset M$ is compact, the closure of $\{x_n(s) : n \geq 2\}$ in M is also compact. Suppose $t \in I \setminus S$. Then $t \in (a, b)$ is a point of continuity of F (with respect to D). By the density of S in I, there is a sequence $\{s'_k\}_{k=1}^{\infty} \subset S \cap (a, t)$ such that $s'_k \to t$ as $k \to \infty$. By (6.4.5), there is an integer $n_1 = n_0(s'_1) \geq 2$ such that $s'_1 = t_{i_1}^{n_1}$ for some $1 \leq i_1 \leq n_1 - 1$. Setting $s = s'_2$ in (6.4.5), choose an integer $n_2 > \max\{n_0(s'_2), n_1\}$, so that $s'_2 = t_{i_2}^{n_2}$ for some $1 \leq i_2 \leq n_2 - 1$. Inductively, if $k \geq 3$, and $n_{k-1} \in \mathbb{N}$ is already chosen, setting $s = s'_k$ in (6.4.5), pick an integer $n_k > \max\{n_0(s'_k), n_{k-1}\}$, so that we have $s'_k = t_{i_k}^{n_k}$ for some $1 \leq i_k \leq n_k - 1$. Since $\{t_i^{n_k}\}_{i=0}^{n_k}$ is a partition of I, there exists a unique integer j_k such that $i_k \leq j_k \leq n_k - 1$ and

$$s'_k = t_{i_k}^{n_k} \leq t_{j_k}^{n_k} < t < t_{j_k+1}^{n_k} \quad \text{for all } k \in \mathbb{N}. \tag{6.4.9}$$

By virtue of definition (6.4.6), $x_{n_k}(t) = x_{j_k}^{n_k} \in F(t_{j_k}^{n_k})$ for all $k \in \mathbb{N}$, and (6.4.9) implies $t_{j_k}^{n_k} \to t$ as $k \to \infty$. Picking, for every $k \in \mathbb{N}$, an element $x_t^{(k)} \in F(t)$ such that $d(x_{j_k}^{n_k}, x_t^{(k)}) = \text{dist}\,(x_{j_k}^{n_k}, F(t))$, by the continuity of F at t and the definition of D, we find

$$d(x_{n_k}(t), x_t^{(k)}) \leq D(F(t_{j_k}^{n_k}), F(t)) \to 0 \quad \text{as} \quad k \to \infty.$$

Since $\{x_t^{(k)}\}_{k=1}^{\infty} \subset F(t)$ and $F(t)$ is compact, there exists a subsequence of $\{x_t^{(k)}\}$, again denoted by $\{x_t^{(k)}\}$, which converges in M to an element $x_t \in F(t)$ as $k \to \infty$, and so,

$$d(x_{n_k}(t), x_t) \leq d(x_{n_k}(t), x_t^{(k)}) + d(x_t^{(k)}, x_t) \to 0 \quad \text{as} \quad k \to \infty.$$

This proves the precompactness of the set $\{x_n(t) : n \in \mathbb{N}\}$.

By Corollary 6.1.7, there is a subsequence of $\{x_n\}$, again denoted by $\{x_{n_k}\}$, which converges pointwise on I to a mapping $x \in BV(I; M)$. It follows from Theorem 6.1.6(b) and (6.4.7) that

$$V(x, I) \leq \liminf_{k \to \infty} V(x_{n_k}, I) \leq V(F, I).$$

Since $x_n(t_0) = x_0$ for $n \geq 2$, we have $x(t_0) = x_0$. The inclusion $x(t) \in F(t)$ for $t \in S$ follows from (6.4.8). If $t \in I \setminus S$, then, by virtue of (6.4.9),

$$\text{dist}\,(x(t), F(t)) \leq d(x(t), x_{n_k}(t)) + \text{dist}\,(x_{n_k}(t), F(t))$$

$$\leq d(x(t), x_{n_k}(t)) + D(F(t_{j_k}^{n_k}), F(t)) \to 0 \quad \text{as } k \to \infty,$$

and so, dist $(x(t), F(t)) = 0$, i.e., $x(t) \in \overline{F(t)} = F(t)$. $\qquad\qquad\square$

6.5 Absolutely Continuous Mappings

An alternative approach to introduce the space $\mathrm{BV}(I; M)$ of mappings $x : I \to M$ of bounded Jordan variation is as follows. In place of (6.1.1) with $\varphi = \mathrm{id}$, consider the pseudomodular w given, for $\lambda > 0$ and $x, y \in X = M^I$, by the rule:

$$w_\lambda(x, y) = \sup \left\{ \sum_{i=1}^n |(x, y)(I_i)| : n \in \mathbb{N} \text{ and } \{I_i\}_1^n \prec I \text{ such that } \lambda \sum_{i=1}^n |I_i| < 1 \right\},$$

$$(6.5.1)$$

where (see Sect. 5.1) if $I_i = \{s_i, t_i\}$, we set $|I_i| = t_i - s_i$. We also put $|I| = b - a$.

Lemma 6.5.1. *For every $\lambda > 0$, w_λ is an extended pseudometric on X and, given $x_0 \in M$, an extended metric on $\dot{X} = \{x : I \to M \mid x(a) = x_0\}$, such that the function $\lambda \mapsto w_\lambda(x, y)$ is nonincreasing and continuous from the right on $(0, \infty)$ for all $x, y \in X$. So, by Sect. 1.3.2, w is a pseudomodular on X and a strict modular on $\dot{X}$. Furthermore, w is nonconvex on X and $\dot{X}$.*

Proof. As usual, in what follows $\lambda, \mu > 0$ and $x, y, z \in X$.

(i) Since $|(x, x)(I_i)| = 0$, we have $w_\lambda(x, x) = \sup\{0\} = 0$. Let us show that if $\lambda > 0$ and $w_\lambda(x, y) = 0$, then $t \mapsto d(x(t), y(t))$ is a constant function on I. By (5.1.10), it suffices to show that $|(x, y)(I_{s,t})| = 0$ for all $I_{s,t} = \{s, t\} \subset I$ with $s < t$. It follows from definition (6.5.1) of w that if $n \in \mathbb{N}$ and

$$\{I_i\}_1^n \prec I \text{ are such that } \lambda \sum_{i=1}^n |I_i| < 1, \text{ then } \sum_{i=1}^n |(x, y)(I_i)| \leq w_\lambda(x, y).$$

$$(6.5.2)$$

Let $\{t_i\}_0^m \prec [s, t]$ be a partition of $[s, t]$ such that $\max_{1 \leq i \leq m}(t_i - t_{i-1}) < 1/\lambda$. (Clearly, if $\lambda < 1/|I|$, then $\lambda(t - s) \leq \lambda|I| < 1$, and so, by (6.5.2), $|(x, y)(I_{s,t})| \leq w_\lambda(x, y) = 0$.) Since $\lambda(t_i - t_{i-1}) < 1$, condition (6.5.2) implies $|(x, y)(I_{t_{i-1}, t_i})| \leq w_\lambda(x, y) = 0$ for all $i = 1, \ldots, m$. By inequality (6.4.2), we get

$$|(x, y)(I_{s,t})| \leq \sum_{i=1}^m |(x, y)(I_{t_{i-1}, t_i})| = 0 \quad \text{for all } s, t \in I,$$

which proves that $t \mapsto d(x(t), y(t))$ is a constant function on I.

Now, if $x, y \in \dot{X}$, then $d(x(t), y(t)) = d(x(a), y(a)) = 0$ for all $t \in I$, and so, $x = y$. This shows that w is strict on $\dot{X}$.

(ii) Because $|(x, y)(I_i)| = |(y, x)(I_i)|$ in (6.5.1), we have $w_\lambda(x, y) = w_\lambda(y, x)$.

(iii) By (5.1.5), for any $n \in \mathbb{N}$ and $\{I_i\}_1^n \prec I$ such that $\lambda \sum_{i=1}^n |I_i| < 1$, we have

$$\sum_{i=1}^n |(x,y)(I_i)| \le \sum_{i=1}^n |(x,z)(I_i)| + \sum_{i=1}^n |(z,y)(I_i)| \le w_\lambda(x,z) + w_\lambda(z,y),$$

which implies $w_\lambda(x,y) \le w_\lambda(x,z) + w_\lambda(z,y)$, and so, w_λ is an extended pseudometric on X and extended metric on $\mathring{X}$.

The function $\lambda \mapsto w_\lambda(x,y)$ is nonincreasing on $(0,\infty)$: if $0 < \lambda_1 < \lambda_2$, $n \in \mathbb{N}$, and $\{I_i\}_1^n \prec I$, then condition $\lambda_2 \sum_{i=1}^n |I_i| < 1$ implies condition $\lambda_1 \sum_{i=1}^n |I_i| < 1$, and so, $w_{\lambda_2}(x,y) \le w_{\lambda_1}(x,y)$. To see that this function is continuous from the right on $(0,\infty)$, we let $n \in \mathbb{N}$ and $\{I_i\}_1^n \prec I$ be such that $\lambda \sum_{i=1}^n |I_i| < 1$. For every $\mu > \lambda$ such that $\mu \sum_{i=1}^n |I_i| < 1$, definition (6.5.1) implies $\sum_{i=1}^n |(x,y)(I_i)| \le w_\mu(x,y)$, and so, passing to the limit as $\mu \to \lambda + 0$, we get $\sum_{i=1}^n |(x,y)(I_i)| \le w_{\lambda+0}(x,y)$. Thus, $w_\lambda(x,y) \le w_{\lambda+0}(x,y)$. The reverse inequality follows from (1.2.4).

In order to prove that w is nonconvex, we first note that

$$w_\lambda(x,y) = \mathrm{w}(x,y) \quad \text{for all} \ \ 0 < \lambda < 1/|I|, \qquad (6.5.3)$$

where w is defined in (6.4.1).

In fact, for every $\{I_i\}_1^n \prec I$, we have $\lambda \sum_{i=1}^n |I_i| \le \lambda|I| < 1$, and so,

$$w_\lambda(x,y) = \sup\left\{ \sum_{i=1}^n |(x,y)(I_i)| : n \in \mathbb{N} \text{ and } \{I_i\}_1^n \prec I \right\}$$

$$= \sup\left\{ \sum_{i=1}^n |(x,y)(I_{t_{i-1},t_i})| : n \in \mathbb{N} \text{ and } \{t_i\}_0^n \prec I \right\} = \mathrm{w}(x,y).$$

Now, the nonconvexity of w follows from Proposition 1.2.3(b) if we note that, for $y = x^\circ$, $x \in BV(I;M)$, $x(a) = x^\circ$, and $x \ne x^\circ$, we have

$$\lim_{\lambda \to +0} w_\lambda(x,x^\circ) = \mathrm{w}(x,x^\circ,I) = V(x,I) < \infty. \qquad \square$$

In order to study the corresponding modular spaces, we prove the following

Lemma 6.5.2. *Given $\lambda > 0$ and $x,y \in X$, we have:*

$$w_\lambda(x,y) \le \mathrm{w}(x,y) \le k(\lambda)w_\lambda(x,y) \quad \text{with} \ \ k(\lambda) = [\lambda|I|] + 1,$$

where $[\alpha]$ denotes the greatest integer not exceeding α (for $\alpha > 0$).

Proof. The left-hand side inequality is clear, because condition '$\{I_i\}_1^n \prec I$ such that $\lambda \sum_{i=1}^n |I_i| < 1$' is more restrictive than simply the condition '$\{I_i\}_1^n \prec I$'.

To establish the right-hand side inequality, let $k \in \mathbb{N}$ and $\{s_j\}_0^k \prec I$ be a partition of I such that $\max_{1 \le j \le k}(s_j - s_{j-1}) < 1/\lambda$. It follows that $|I| = \sum_{j=1}^k (s_j - s_{j-1}) < k/\lambda$, which implies $k > \lambda|I|$, i.e., $k \ge k(\lambda)$. Given $1 \le j \le k$, for any $n \in \mathbb{N}$ and a partition $\{t_i\}_0^n \prec [s_{j-1}, s_j]$, setting $I_i = \{t_{i-1}, t_i\}$ for $i = 1, \ldots, n$, we find $\{I_i\}_1^n \prec I$ and

$$\lambda \sum_{i=1}^n |I_i| = \lambda \sum_{i=1}^n (t_i - t_{i-1}) = \lambda(s_j - s_{j-1}) < 1.$$

By (6.5.2), $\sum_{i=1}^n |(x,y)(I_i)| \le w_\lambda(x,y)$, and the arbitrariness of n and $\{t_i\}_0^n$ imply $w(x, y, [s_{j-1}, s_j]) \le w_\lambda(x,y)$ for all $j = 1, \ldots, k$. By the additivity property of w (Lemma 6.4.1), we conclude that

$$w(x, y) \equiv w(x, y, I) = \sum_{j=1}^k w(x, y, [s_{j-1}, s_j]) \le k w_\lambda(x, y),$$

and the right-hand side inequality follows with $k = k(\lambda)$. $\square$

Remark 6.5.3. If $0 < \lambda < 1/|I|$ in Lemma 6.5.2, then $k(\lambda) = [\lambda|I|] + 1 = 1$, and so, $w_\lambda(x, y) = w(x, y)$, which is exactly equality (6.5.3).

Recall that a mapping $x : I \to M$ is said to be *absolutely continuous* (in symbols, $x \in AC(I; M)$) if for every $\varepsilon > 0$ there is $\delta(\varepsilon) > 0$ such that

$$\sum_{i=1}^n |x(I_i)| < \varepsilon \text{ for all } n \in \mathbb{N} \text{ and } \{I_i\}_1^n \prec I \text{ with } \sum_{i=1}^n |I_i| < \delta(\varepsilon). \qquad (6.5.4)$$

Theorem 6.5.4. $X_w^0 = AC(I; M) \subset X_w^* = X_w^{\text{fin}} = BV(I; M)$.

Proof. (Recall that $x^\circ(t) = x^\circ$ for all $t \in I$, where $x^\circ \in M$ is fixed.)

1. Let us show that $BV(I; M) \subset X_w^{\text{fin}}$. If $x \in BV(I; M)$, then $V(x, I) = w(x, x^\circ) < \infty$, and so, by Lemma 6.5.2, we have $w_\lambda(x, x^\circ) \le w(x, x^\circ) < \infty$ for all $\lambda > 0$, which means that $x \in X_w^{\text{fin}} \subset X_w^*$. Now, we show that $X_w^* \subset BV(I; M)$. In fact, if $x \in X_w^*$, then $w_\lambda(x, x^\circ)$ is finite for some $\lambda > 0$. Applying Lemma 6.5.2 once again, we get $w(x, x^\circ) \le k(\lambda) w_\lambda(x, x^\circ) < \infty$, i.e., $x \in BV(I; M)$.

2. We know from Sect. 2.1 that $X_w^0 \subset X_w^*$. To see that $X_w^0 \subset AC(I; M)$, let $x \in X_w^0$, i.e., $w_\lambda(x, x^\circ) \to 0$ as $\lambda \to \infty$. This implies that for every $\varepsilon > 0$ there is $\lambda_0(\varepsilon) > 0$ such that $w_\lambda(x, x^\circ) < \varepsilon$ for all $\lambda \ge \lambda_0(\varepsilon)$. Setting $\delta(\varepsilon) = 1/\lambda_0(\varepsilon)$, we find $w_{1/\delta(\varepsilon)}(x, x^\circ) < \varepsilon$. By definition (6.5.1), this means that condition (6.5.4) holds, and so, $x \in AC(I; M)$. The reverse inclusion $AC(I; M) \subset X_w^0$ is obtained by reversing the previous arguments. In fact, if $x \in AC(I; M)$, then for each $\varepsilon > 0$ there is $\delta(\varepsilon) > 0$ such that (6.5.4) holds. Definition (6.5.1)

implies $w_{1/\delta(\varepsilon)}(x, x^\circ) \le \varepsilon$. By Lemma 6.5.1, the function $\lambda \mapsto w_\lambda(x, x^\circ)$ is nonincreasing, and so, if $\lambda \ge 1/\delta(\varepsilon)$, $w_\lambda(x, x^\circ) \le w_{1/\delta(\varepsilon)}(x, x^\circ) \le \varepsilon$. This yields $\lim_{\lambda \to \infty} w_\lambda(x, x^\circ) = 0$ and $x \in X_w^0$.   □

Remark 6.5.5. The pseudometric $d_w^0(x, y) = \inf\{\lambda > 0 : w_\lambda(x, y) \le \lambda\}$ on the modular space $X_w^* = BV(I; M)$ satisfies the inequalities:

$$\frac{\sqrt{1 + 4|I|\mathrm{w}(x, y)} - 1}{2|I|} \le d_w^0(x, y) \le \mathrm{w}(x, y).$$

In order to see this, denote by λ_0 the quantity at left-hand side and note that it is the nonnegative root of the quadratic equation $|I|\lambda^2 + \lambda - \mathrm{w}(x, y) = 0$. Now, if $\lambda > 0$ is such that $w_\lambda(x, y) \le \lambda$, then, by Lemma 6.5.2, we get

$$\mathrm{w}(x, y) \le k(\lambda)w_\lambda(x, y) \le ([\lambda|I|] + 1)\lambda \le \lambda^2|I| + \lambda,$$

and so, $\lambda_0 \le \lambda$, which establishes the left-hand side inequality. For the right-hand side inequality, we have: if $\lambda > 0$ and $\lambda \ge \mathrm{w}(x, y)$, then Lemma 6.5.2 implies $w_\lambda(x, y) \le \mathrm{w}(x, y) \le \lambda$, and so, by the definition of d_w^0, $d_w^0(x, y) \le \mathrm{w}(x, y)$.

The above inequalities and Lemma 6.5.2 imply that the d_w^0-convergence and modular convergence (with respect to w from (6.5.1)) are equivalent on $X_w^* = BV(I; M)$.

6.6 The Riesz-Medvedev Generalized Variation

In this section, $I = [a, b]$, $|I| = b - a$, (M, d) is a metric space, $X = M^I$, and Φ is a *convex* function on $[0, \infty)$ such that $\Phi(u) = 0$ iff $u = 0$ (see Appendix A.1), and so,

$$[\Phi] \equiv \lim_{u \to \infty} \Phi(u)/u = \sup_{u > 0} \Phi(u)/u \in (0, \infty].$$

The inverse function Φ^{-1} of Φ is increasing, continuous, and concave on $[0, \infty)$. Moreover, the function $\omega_\Phi(u) = u\,\Phi^{-1}(1/u)$, $u > 0$, is nondecreasing, continuous, and concave (hence subadditive), and

$$\omega_\Phi(0) \equiv \omega_\Phi(+0) = \lim_{u \to +0} \omega_\Phi(u) = 1/[\Phi] \in [0, \infty).$$

Given $\lambda > 0$ and $x, y \in X$, changing the variational core in (6.1.1), we set

$$w_\lambda(x, y) \equiv w_\lambda^\Phi(x, y, I) = \sup\left\{\sum_{i=1}^n |I_i|\Phi\left(\frac{|(x, y)(I_i)|}{\lambda|I_i|}\right) : n \in \mathbb{N} \text{ and } \{t_i\}_0^n \prec I\right\},$$

$$\tag{6.6.1}$$

where $I_i = \{t_{i-1}, t_i\}$, $|I_i| = t_i - t_{i-1}$, $\{t_i\}_0^n$ is a partition of I, and the joint increment $|(x, y)(I_i)|$ is given by (5.1.1) (or in particular cases by (5.1.2) or (5.1.3)).

Note that Lemma 6.1.2 holds for the function w from (6.6.1). In this case, inequality (6.1.2) is replaced by the inequality

$$(t-s)\Phi\left(\frac{|(x,y)(I_{s,t})|}{\lambda(t-s)}\right) \le w_\lambda(x,y), \qquad I_{s,t} = \{s,t\} \subset I, \quad s < t,$$

and so,

$$|(x,y)(I_{s,t})| \le \lambda|t-s|\Phi^{-1}\left(\frac{w_\lambda(x,y)}{|t-s|}\right), \qquad s,t \in I, \quad s \ne t. \tag{6.6.2}$$

If, as usual, $x^\circ(t) \equiv x^\circ$ on I with $x^\circ \in M$, we find that the quantity

$$w_\lambda(x,x^\circ) = \sup\left\{\sum_{i=1}^n |I_i|\Phi\left(\frac{|x(I_i)|}{\lambda|I_i|}\right) : n \in \mathbb{N} \text{ and } \{t_i\}_0^n \prec I\right\}$$

is independent of the constant mapping x°.

The quantity $V_\Phi(x,I) = w_1(x,x^\circ) \equiv w_1^\Phi(x,x^\circ,I)$ is known as the generalized Φ-*variation* of $x \in X$ on I in the sense of F. Riesz and Yu. T. Medvedev, and x is said to be *of bounded Φ-variation* if $V_\Phi(x,I) < \infty$. Clearly, $w_\lambda(x,x^\circ) = V_{\Phi_\lambda}(x,I)$, where $\Phi_\lambda(u) = \Phi(u/\lambda)$ for $u \ge 0$ and $\lambda > 0$.

We denote by $\mathrm{GV}_\Phi(I;M) = X_w^* = X_w^*(x^\circ)$ the modular space of all mappings $x : I \to M$ of bounded generalized Φ-variation on I. Since Φ is convex, by Lemma 6.1.2, we have $X_w^0 = X_w^*$.

Let $\mathrm{Lip}(I;M)$ be the space of all Lipschitzian mappings $x : I \to M$ (it is introduced by setting $N = I$ with natural metric in Sect. 6.2). The least Lipschitz contant (6.2.2) of $x \in \mathrm{Lip}(I;M)$ will be denoted by $L(x,I) = \sup\{d(x(t),x(s))/|t-s| : t,s \in I, t \ne s\}$.

Theorem 6.6.1. $\mathrm{Lip}(I;M) \subset X_w^{\mathrm{fin}} \subset X_w^* = \mathrm{GV}_\Phi(I;M) \subset \mathrm{BV}(I;M)$; *moreover,*

$$X_w^{\mathrm{fin}} = \mathrm{GV}_\Phi(I;M) = \mathrm{BV}(I;M) \text{ if } [\Phi] < \infty, \text{ and} \tag{6.6.3}$$

$$\mathrm{GV}_\Phi(I;M) \subset \mathrm{AC}(I;M) \text{ if } [\Phi] = \infty. \tag{6.6.4}$$

Proof. 1. Let $x \in \mathrm{Lip}(I;M)$. We have $|x(I_i)| \le L(x,I_i)|I_i|$ for any partition $\{t_i\}_0^n \prec I$ with $I_i = \{t_{i-1},t_i\}$, and so, by the monotonicity of Φ,

$$\sum_{i=1}^n |I_i|\Phi\left(\frac{|x(I_i)|}{\lambda|I_i|}\right) \le \sum_{i=1}^n |I_i|\Phi\left(\frac{L(x,I_i)}{\lambda}\right) \le |I|\Phi(L(x,I)/\lambda).$$

Taking the supremum over all partitions of I, we get

$$w_\lambda(x,x^\circ) = V_{\Phi_\lambda}(x,I) \le |I|\Phi(L(x,I)/\lambda), \qquad \lambda > 0,$$

which implies $x \in X_w^{\mathrm{fin}} \subset X_w^* = \mathrm{GV}_\Phi(I;M)$.

2. To show that $\mathrm{GV}_\Phi(I;M) \subset \mathrm{BV}(I;M)$, we argue as follows. If $x,y \in X$ are such that $x \sim y$, then $w_\lambda(x,y) < \infty$ for some $\lambda = \lambda(x,y) > 0$. Given $\{I_i\}_1^n \prec I$, setting $\alpha_i = |I_i|$ and $u_i = |(x,y)(I_i)|/\lambda|I_i|$ in Jensen's inequality (A.1.2), we have

$$\Phi\left(\frac{\sum_{i=1}^{n}|(x,y)(I_i)|}{\lambda\sum_{i=1}^{n}|I_i|}\right) \le \frac{1}{\sum_{i=1}^{n}|I_i|}\cdot\sum_{i=1}^{n}|I_i|\Phi\left(\frac{|(x,y)(I_i)|}{\lambda|I_i|}\right) \le \frac{w_\lambda(x,y)}{\sum_{i=1}^{n}|I_i|},$$

and so, taking the inverse function Φ^{-1}, we find

$$\sum_{i=1}^{n}|(x,y)(I_i)| \le \lambda\left(\sum_{i=1}^{n}|I_i|\right)\Phi^{-1}\left(\frac{w_\lambda(x,y)}{\sum_{i=1}^{n}|I_i|}\right). \tag{6.6.5}$$

Now, if $\{t_i\}_0^n \prec I$ is a partition of I with $I_i = \{t_{i-1},t_i\}$, then $\sum_{i=1}^{n}|I_i| = |I|$, and it follows from (6.4.1) and (6.6.5) that

$$\mathrm{w}(x,y) \equiv \mathrm{w}(x,y,I) \le \lambda|I|\Phi^{-1}\big(w_\lambda(x,y,I)/|I|\big). \tag{6.6.6}$$

Assuming that $x \in \mathrm{GV}_\Phi(I;M)$ and $y = x^\circ$ in (6.6.6), we get $V(x,I) = \mathrm{w}(x,x^\circ) < \infty$, i.e., $x \in \mathrm{BV}(I;M)$.

3. Let us prove (6.6.3). It suffices to show that $\mathrm{BV}(I;M) \subset X_w^{\mathrm{fin}}$. If $x \in \mathrm{BV}(I;M)$ and $\{I_i\}_1^n \prec I$, then taking into account the inequality $\Phi(u) \le [\Phi]u$ for all $u \ge 0$, we find

$$\sum_{i=1}^{n}|I_i|\Phi\left(\frac{|x(I_i)|}{\lambda|I_i|}\right) \le \frac{[\Phi]}{\lambda}\sum_{i=1}^{n}|x(I_i)| \le \frac{[\Phi]}{\lambda}V(x,I).$$

In particular, if $\sum_{i=1}^{n}|I_i| = |I|$, this implies

$$w_\lambda(x,x^\circ) \le \frac{[\Phi]}{\lambda}V(x,I) < \infty \quad \text{for all} \quad \lambda > 0,$$

and so, $x \in X_w^{\mathrm{fin}}$.

4. In order to establish (6.6.4), let $x \in \mathrm{GV}_\Phi(I;M)$, so that $w_\lambda(x,x^\circ)$ is finite for some $\lambda > 0$. If $w_\lambda(x,x^\circ) = 0$, then (6.6.2) implies $d(x(t),x(s)) = |(x,x^\circ)(I_{s,t})| = 0$ for all $s,t \in I$, i.e., x is a constant mapping on I. Assuming that $w_\lambda(x,x^\circ) \ne 0$, let $\{I_i\}_1^n \prec I$. Since $[\Phi] = \infty$, we find

$$\lim_{r\to+0} r\,\Phi^{-1}(w_\lambda(x,x^\circ)/r) = w_\lambda(x,x^\circ)\lim_{u\to\infty}\frac{u}{\Phi(u)} = 0.$$

Hence, given $\varepsilon > 0$ there exists $\delta(\varepsilon) > 0$ such that $r\,\Phi^{-1}(w_\lambda(x,x^\circ)/r) < \varepsilon/\lambda$ for all $0 < r < \delta(\varepsilon)$. This and inequality (6.6.5) with $y = x^\circ$ yield condition (6.5.4), and so, $x \in \mathrm{AC}(I;M)$. $\qquad\square$

Due to (6.6.3), in the sequel we mainly study the case when $[\Phi] = \infty$.

As the following lemma shows, quantity (6.6.1) is additive in the third variable.

Lemma 6.6.2. *Given* $x,y \in X$, $a \le s < t \le b$, *and* $\lambda > 0$, *we have*

$$w_\lambda(x,y,[a,s]) + w_\lambda(x,y,[s,t]) = w_\lambda(x,y,[a,t]).$$

Proof. Inequality ($\leq$) is established as in Theorem 6.1.6(a), so we concentrate on the reverse inequality. Let $\{t_i\}_0^n \prec [a, t]$ be a partition of $[a, t]$. With no loss of generality, we may assume that $t_{k-1} < s < t_k$ for some $1 \leq k \leq n$. Setting $I' = I_{t_{k-1}, s} = \{t_{k-1}, s\}$ and $I'' = I_{s, t_k} = \{s, t_k\}$, we find $|I_k| = |I'| + |I''|$ and, by virtue of (6.4.2), $|(x, y)(I_k)| \leq |(x, y)(I')| + |(x, y)(I'')|$. From the convexity (and monotonicity) of Φ, we have

$$|I_k|\Phi\left(\frac{|(x, y)(I_k)|}{\lambda|I_k|}\right) \leq |I'|\Phi\left(\frac{|(x, y)(I')|}{\lambda|I'|}\right) + |I''|\Phi\left(\frac{|(x, y)(I'')|}{\lambda|I''|}\right).$$

Applying this inequality to the k-th term in the sum in (6.6.1) corresponding to $I = [a, t]$ and noting that $\{t_i\}_0^{k-1} \cup \{s\}$ is a partition of $[a, s]$ and $\{s\} \cup \{t_i\}_k^n$ is a partition of $[s, t]$, we obtain the desired inequality ($\geq$). $\qquad\square$

The structure of mappings from $\mathrm{GV}_\Phi(I; M)$ is clarified in the following

Theorem 6.6.3. *Given $x \in \mathrm{BV}(I; M)$, set $v(t) = V(x, [a, t])$ for $t \in I$, and let $J = v(I)$ be the image of I under v. We have: $x \in \mathrm{GV}_\Phi(I; M)$ if and only if $v \in \mathrm{GV}_\Phi(I; \mathbb{R})$ and there is $x_1 \in \mathrm{Lip}(J; M)$ such that $L(x_1, J) \leq 1$ and $x(t) = x_1(v(t))$ for all $t \in I$. Furthermore, $V_{\Phi_\lambda}(v, I) = w_\lambda(x, x^\circ, I) = V_{\Phi_\lambda}(x, I)$ in $[0, \infty]$ for all $\lambda > 0$.*

Proof. ($\Rightarrow$) Suppose $x \in \mathrm{GV}_\Phi(I; M)$. Then $V_{\Phi_\lambda}(x, I) = w_\lambda(x, x^\circ) < \infty$ for some $\lambda > 0$. By Theorem 6.6.1 and inequality (6.6.6) (with $y = x^\circ$), the function v is well-defined, bounded, nondecreasing, and (absolutely) continuous on I. Hence $J = [0, \ell]$ where $\ell = V(x, I)$. Given a partition $\{t_i\}_0^n \prec I$ with $I_i = \{t_{i-1}, t_i\}$ and $|I_i| = t_i - t_{i-1}$, the additivity of w (Lemma 6.4.1) and inequality (6.6.6) imply

$$|v(I_i)| = v(t_i) - v(t_{i-1}) = w(x, x^\circ, [t_{i-1}, t_i]) \leq \lambda|I_i|\Phi^{-1}\left(\frac{w_\lambda(x, x^\circ, [t_{i-1}, t_i])}{|I_i|}\right).$$

From the monotonicity of Φ and Lemma 6.6.2, we get

$$\sum_{i=1}^n |I_i|\Phi\left(\frac{|v(I_i)|}{\lambda|I_i|}\right) \leq \sum_{i=1}^n w_\lambda(x, x^\circ, [t_{i-1}, t_i]) = w_\lambda(x, x^\circ, I) = V_{\Phi_\lambda}(x, I).$$

This implies $v \in \mathrm{GV}_\Phi(I; \mathbb{R})$ along with the inequality

$$V_{\Phi_\lambda}(v, I) \leq V_{\Phi_\lambda}(x, I). \tag{6.6.7}$$

For $\tau \in J = [0, \ell]$, we set $x_1(\tau) = x(t)$ where $t \in I$ is any element such that $\tau = v(t)$ (more explicitly, $x_1(v(t)) = x(t)$). The mapping $x_1 : J \to M$ is well-defined: by virtue of (6.4.3), if $s \in I$ and $v(s) = \tau$, then $x(s) = x(t)$. In order to see that $L(x_1, J) \leq 1$, we let $\tau, \sigma \in J$, so that $\tau = v(t)$ and $\sigma = v(s)$ for some $t, s \in I$. It follows from (6.4.3) that

$$d(x_1(\tau), x_1(\sigma)) = d(x(t), x(s)) \leq |v(t) - v(s)| = |\tau - \sigma|.$$

($\Leftarrow$) Since $v \in GV_\Phi(I; \mathbb{R})$, we have $V_{\Phi_\lambda}(v, I) < \infty$ for some $\lambda > 0$, and by (6.6.4), $v \in AC(I; \mathbb{R})$ (and so, $J = v(I)$ is a closed interval in $\mathbb{R}$). For any partition $\{t_i\}_0^n \prec I$ of I with $I_i = \{t_{i-1}, t_i\}$, we find

$$\sum_{i=1}^n |I_i| \Phi\left(\frac{|x(I_i)|}{\lambda |I_i|}\right) = \sum_{i=1}^n |I_i| \Phi\left(\frac{d(x_1(v(t_i)), x_1(v(t_{i-1})))}{\lambda |I_i|}\right)$$

$$\leq \sum_{i=1}^n |I_i| \Phi\left(L(x_1, J) \frac{|v(I_i)|}{\lambda |I_i|}\right) \leq V_{\Phi_\lambda}(v, I),$$

and so, $V_{\Phi_\lambda}(x, I) \leq V_{\Phi_\lambda}(v, I)$. It remains to take into account inequality (6.6.7). $\quad\square$

The quantity (6.6.1) with $y = x^\circ$ allows the integral representation as follows.

Theorem 6.6.4. *Let $(M, \|\cdot\|)$ be a reflexive Banach space and $x \in GV_\Phi(I; M)$. Then x is norm-differentiable almost everywhere on I, its derivative x' is strongly measurable and Bochner integrable on I, $x(t) = x(a) + \int_a^t x'(\tau)\,d\tau$ for all $t \in I$, and*

$$V_{\Phi_\lambda}(x, I) = V_\Phi(x/\lambda, I) = \int_I \Phi\left(\frac{\|x'(t)\|}{\lambda}\right) dt, \quad \lambda > 0. \tag{6.6.8}$$

Proof. By the assumption, $V_{\Phi_\lambda}(x, I) = w_\lambda(x, x^\circ, I) < \infty$ for some $\lambda > 0$.

1. First, we show that if $0 < s < |I| = b - a$, then

$$\int_a^{b-s} \Phi\left(\frac{\|x(t+s) - x(t)\|}{\lambda s}\right) dt \leq V_{\Phi_\lambda}(x, I). \tag{6.6.9}$$

By Lemma 6.6.2, function $t \mapsto V_{\Phi_\lambda}(x, [a, t])$ is nondecreasing on I and, hence, Riemann integrable. Since, by (6.6.4), $x \in AC(I; M)$, $t \mapsto \|x(t+s) - x(t)\|$ is a continuous function for $t \in [a, b-s]$. From the definition of the Φ_λ-variation and Lemma 6.6.2, we have

$$\Phi\left(\frac{\|x(t+s) - x(t)\|}{\lambda s}\right) \leq \frac{1}{s}\left(V_{\Phi_\lambda}(x, [a, t+s]) - V_{\Phi_\lambda}(x, [a, t])\right).x$$

Integrating over $t \in [a, b - s]$ and changing variables appropriately, we get

$$\int_a^{b-s} \Phi\left(\frac{\|x(t+s) - x(t)\|}{\lambda s}\right) dt \leq \frac{1}{s} \int_{b-s}^b V_{\Phi_\lambda}(x, [a, t])\,dt \leq V_{\Phi_\lambda}(x, I).$$

2. Since x is absolutely continuous on I, all assertions of the theorem, except (6.6.8), are established in a standard way (see comments in Sect. 6.7). In order to obtain (6.6.8), we note that

$$\|x'(t)\| \leq \liminf_{s \to +0} \frac{\|x(t+s) - x(t)\|}{s} \quad \text{for almost every } t \in I,$$

and so, inequality (6.6.9) and Fatou's Lemma imply

$$\int_I \Phi\left(\frac{\|x'(t)\|}{\lambda}\right) dt \leq \liminf_{s \to +0} \int_a^{b-s} \Phi\left(\frac{\|x(t+s) - x(t)\|}{\lambda s}\right) dt \leq V_{\Phi_\lambda}(x, I).$$

The reverse inequality follows from the integral representation of x and Jensen's integral inequality (A.1.3) (see Appendix A.1): in fact, if $\{t_i\}_0^n$ is a partition of I and $I_i = \{t_{i-1}, t_i\}$, we have

$$\sum_{i=1}^n |I_i| \Phi\left(\frac{|x(I_i)|}{\lambda |I_i|}\right) = \sum_{i=1}^n (t_i - t_{i-1}) \Phi\left(\frac{\|x(t_i) - x(t_{i-1})\|}{\lambda(t_i - t_{i-1})}\right)$$

$$\leq \sum_{i=1}^n (t_i - t_{i-1}) \Phi\left(\frac{1}{t_i - t_{i-1}} \int_{t_{i-1}}^{t_i} \frac{\|x'(t)\|}{\lambda} dt\right)$$

$$\leq \sum_{i=1}^n \int_{t_{i-1}}^{t_i} \Phi\left(\frac{\|x'(t)\|}{\lambda}\right) dt = \int_I \Phi\left(\frac{\|x'(t)\|}{\lambda}\right) dt.$$

Thus $V_{\Phi_\lambda}(x, I) \leq \int_I \Phi(\|x'(t)\|/\lambda) dt$, which completes the proof.   □

Corollary 6.6.5. (a) *If $(M, \|\cdot\|)$ is a reflexive Banach space, then $x \in GV_\Phi(I; M)$ if and only if $x \in AC(I; M)$ and $\int_I \Phi(\|x'(t)\|/\lambda) dt < \infty$ for some $\lambda > 0$.*
(b) *If (M, d) is a metric space, $x : I \to M$, and $v(t) = V(x, [a, t])$ for $t \in I$, then $x \in GV_\Phi(I; M)$ if and only if $v \in AC(I; \mathbb{R})$ and $\int_I \Phi(|v'(t)|/\lambda) dt < \infty$ for some $\lambda > 0$, in which case we have*

$$V_{\Phi_\lambda}(x, I) = V_\Phi(v/\lambda, I) = \int_I \Phi\left(\frac{1}{\lambda} \cdot \frac{d}{dt} V(x, [a, t])\right) dt.$$

Proof. (a) is a straightforward consequence of Theorem 6.6.4.
(b) follows from item (a), the reflexivity of $\mathbb{R}$, and Theorem 6.6.3.   □

Remark 6.6.6. (1) The interest in mappings of bounded generalized Riesz-Medvedev Φ-variation lies in the fact that only such mappings are absolutely continuous on I: any $x \in AC(I; M)$ belongs to some $GV_\Phi(I; M)$. In fact, since $v(t) = V(x, [a, t])$, $t \in I$, is in $AC(I; \mathbb{R})$, Lebegue's Theorem on the differentiation of numerical functions implies that the (almost everywhere) derivative v' is Lebesgue integrable on I. It follows from Krasnosel'skiĭ and Rutickiĭ [56, Sect. 8.1] that there exists a convex φ-function Φ on $[0, \infty)$ with $[\Phi] = \infty$ such that $\int_I \Phi(|v'(t)|) dt < \infty$. Corollary 6.6.5(b) yields $x \in GV_\Phi(I; M)$.
(2) Similar to Lemma 6.1.5(b), given $x, y \in GV_\Phi(I; M)$, the function $d_M(x, y) = d(x(a), y(a)) + d_w^*(x, y)$ is a metric on $GV_\Phi(I; M)$, for which we have the inequality $d_\infty(x, y) \leq \max\{1, \omega_\Phi(|I|)\} d_M(x, y)$. In fact, for any $\lambda > d_w^*(x, y)$,

we get $w_\lambda(x, y) \leq 1$, and so, inequality (6.6.6) implies $\mathrm{w}(x, y) \leq \lambda\omega_\Phi(|I|)$. Thus $\mathrm{w}(x, y) \leq d_w^*(x, y)\omega_\Phi(|I|)$, and it follows from (5.1.10) that

$$d(x(t), y(t)) \leq d(x(a), y(a)) + |(x, y)(I_{a,t})| \leq d(x(a), y(a)) + \mathrm{w}(x, y)$$

$$\leq d(x(a), y(a)) + d_w^*(x, y)\omega_\Phi(|I|) \quad \text{for all} \quad t \in I.$$

(3) Note that the d_w^*-convergence and w-convergence (see (6.6.6)) on the space $\dot{X}_w^* = \{x \in GV_\Phi(I; M) : x(a) = x_0\}$ imply the uniform convergence.

(4) It follows from (6.6.2) that $x \in GV_\Phi(I; M)$ has the "qualified" modulus of continuity of the form $d(x(t), x(s)) \leq d_w^*(x, x^\circ)\omega_\Phi(|t - s|)$ for $t, s \in I$.

(5) Counterparts of Theorem 6.1.6(b), (c) also hold in the space $GV_\Phi(I; M)$. An analogue of Theorem 6.3.1 holds if we replace BV_φ by GV_Φ, and $\gamma(\varphi)$—by the quantity $\gamma(\Phi) = \max\{1, 2\omega_\Phi(|I|)\}$.

Example 6.6.7. Here we present an example when the metric and modular convergences coincide in the space $GV_\Phi(I; M)$. In order to be able to calculate explicitly, we set $I = [0, 1]$, $M = \mathbb{R}$, and $\Phi(u) = e^u - 1$ for $u \geq 0$. Clearly, $[\Phi] = \infty$, and Φ does not satisfy the Δ_2-condition (see Example 4.2.7(1)). For $\alpha > 0$, define $x_\alpha : [0, 1] \to \mathbb{R}$ by

$$x_\alpha(t) = \alpha t(1 - \log t) \quad \text{if} \quad 0 < t \leq 1, \quad \text{and} \quad x_\alpha(0) = 0.$$

Since $x_\alpha'(t) = -\alpha \log t$ for $0 < t \leq 1$, by virtue of (6.6.8), we have

$$w_\lambda(x_\alpha, 0) = \int_0^1 \Phi\left(\frac{|x_\alpha'(t)|}{\lambda}\right) dt = \int_0^1 \frac{dt}{t^{\alpha/\lambda}} - 1 = \begin{cases} \infty & \text{if } 0 < \lambda \leq \alpha, \\ \dfrac{\alpha}{\lambda - \alpha} & \text{if} \quad \lambda > \alpha. \end{cases}$$

Hence $x_\alpha \in X_w^*(0) = GV_\Phi(I; M)$ for all $\alpha > 0$. Noting that

$$d_w^*(x_\alpha, 0) = \inf\{\lambda > 0 : w_\lambda(x_\alpha, 0) \leq 1\} = 2\alpha,$$

we find that, as $\alpha \to +0$, $d_w^*(x_\alpha, 0) \to 0$ and $w_\lambda(x_\alpha, 0) \to 0$ for all $\lambda > 0$, and (in accordance with Theorem 4.1.1) these two convergences are equivalent.

Example 6.6.8. Here we exhibit an example when the modular convergence in $GV_\Phi(I; M)$ is weaker than the metric convergence. Let I, M, and Φ be as in the previous Example 6.6.7. Given $0 \leq \beta \leq 1$, define $x_\beta : [0, 1] \to \mathbb{R}$ as follows:

$$x_\beta(t) = t - (t + \beta)\log(t + \beta) + \beta \log \beta \quad \text{if} \quad \beta > 0 \quad \text{and} \quad 0 \leq t \leq 1$$

and

$$x_0(t) = t - t\log t \quad \text{if} \quad 0 < t \leq 1, \quad \text{and} \quad x_0(0) = 0.$$

Since $x'_\beta(t) = -\log(t + \beta)$ for $\beta > 0$ and $t \in [0, 1]$, we have

$$|x'_\beta(t)| = -\log(t+\beta) \quad \text{if} \quad 0 \leq t \leq 1-\beta, \quad \text{and} \quad |x'_\beta(t)| = \log(t+\beta) \quad \text{if} \quad 1-\beta < t \leq 1,$$

and so, applying the integral formula (6.6.8), we find

$$w_\lambda(x_\beta, 0) = \int_0^1 \Phi\left(\frac{|x'_\beta(t)|}{\lambda}\right) dt = I_1 + I_2 - 1, \qquad \lambda, \beta > 0,$$

where

$$I_1 = \int_0^{1-\beta} \frac{dt}{(t+\beta)^{1/\lambda}} = \begin{cases} \dfrac{\lambda}{\lambda - 1}\left(1 - \beta^{(\lambda-1)/\lambda}\right) & \text{if} \quad 0 < \lambda \neq 1, \\ -\log\beta & \text{if} \qquad \lambda = 1, \end{cases}$$

and

$$I_2 = \int_{1-\beta}^1 (t+\beta)^{1/\lambda}\, dt = \frac{\lambda}{\lambda+1}\left((1+\beta)^{(\lambda+1)/\lambda} - 1\right) \quad \text{for all} \quad \lambda > 0.$$

From Example 6.6.7 with $\alpha = 1$, we also have $w_\lambda(x_0, 0) = \infty$ if $0 < \lambda \leq 1$, and $w_\lambda(x_0, 0) = 1/(\lambda - 1)$ if $\lambda > 1$. Thus $x_\beta \in X_w^*(0) = \mathrm{GV}_\Phi(I; M)$ for all $0 \leq \beta \leq 1$.

Clearly, x_β converges pointwise on $[0, 1]$ to x_0 as $\beta \to +0$.

Let us calculate the values $w_\lambda(x_\beta, x_0)$ for $\lambda > 0$ and $d_w^*(x_\beta, x_0)$ and investigate their convergence to zero as $\beta \to +0$. Since

$$(x_\beta - x_0)'(t) = -\log(t+\beta) + \log t \quad \text{for} \quad 0 < t \leq 1,$$

we have

$$\frac{|(x_\beta - x_0)'(t)|}{\lambda} = \frac{\log(t+\beta) - \log t}{\lambda} = \log\left(1 + \frac{\beta}{t}\right)^{1/\lambda},$$

and so, applying (6.6.8), we get

$$w_\lambda(x_\beta, x_0) = w_\lambda(x_\beta - x_0, 0) = \int_0^1 \Phi\left(\frac{|(x_\beta - x_0)'(t)|}{\lambda}\right) dt = -1 + \int_0^1 \left(1 + \frac{\beta}{t}\right)^{1/\lambda} dt.$$

If $0 < \lambda \leq 1$, then

$$\left(1 + \frac{\beta}{t}\right)^{1/\lambda} \geq 1 + \frac{\beta}{t} \quad \text{and} \quad \int_0^1 \left(1 + \frac{\beta}{t}\right) dt = \infty,$$

and so, $w_\lambda(x_\beta, x_0) = \infty$ for all $0 < \beta \leq 1$ and $0 < \lambda \leq 1$.

Suppose $\lambda > 1$. We have

$$w_\lambda(x_\beta, x_0) = -1 + \int_0^\beta \left(1 + \frac{\beta}{t}\right)^{1/\lambda} dt + \int_\beta^1 \left(1 + \frac{\beta}{t}\right)^{1/\lambda} dt \equiv -1 + \mathit{II}_1 + \mathit{II}_2,$$

where

$$\mathit{II}_1 \leq \int_0^\beta \left(\frac{2\beta}{t}\right)^{1/\lambda} dt = (2\beta)^{1/\lambda} \int_0^\beta t^{-1/\lambda} dt = (2\beta)^{1/\lambda} \cdot \frac{\lambda}{\lambda - 1} \cdot \beta^{1-(1/\lambda)} =$$

$$= 2^{1/\lambda} \cdot \frac{\lambda\beta}{\lambda - 1} \to 0 \quad \text{as} \quad \beta \to +0$$

and

$$\mathit{II}_2 \leq \int_\beta^1 \left(1 + \frac{\beta}{t}\right) dt = (1 - \beta) - \beta \log \beta \to 1 \quad \text{as} \quad \beta \to +0.$$

It follows that $w_\lambda(x_\beta, x_0) \to 0$ as $\beta \to +0$ for all $\lambda > 1$.

On the other hand, $w_\lambda(x_\beta, x_0) = \infty$ for all $0 < \beta \leq 1$ and $0 < \lambda \leq 1$, and so,

$$d_w^*(x_\beta, x_0) = \inf\{\lambda > 0 : w_\lambda(x_\beta, x_0) \leq 1\} \geq 1,$$

and $d_w^*(x_\beta, x_0)$ does not converge to zero as $\beta \to +0$.

Next we are going to show that the space $GV_\Phi(I; M)$ is adequate for solving certain Carathéodory-type ordinary differential equations.

Let $(M, \|\cdot\|)$ be a reflexive Banach space. We denote by $L_1(I; M)$ the space of all strongly measurable and Bochner integrable mappings $x : I \to M$, and by $L_\Phi(I; M)$—the Orlicz space of all strongly measurable mappings $x : I \to M$ such that $\rho(\alpha x) < \infty$ for some constant $\alpha = \alpha(x) > 0$, where $\rho(x) = \int_I \Phi(\|x(t)\|) dt$ is the classical Orlicz convex modular (see Musielak [75, Chap. II]).

Given $x_0 \in M$, we set $\dot{X}_w^* = \{x \in GV_\Phi(I; M) : x(a) = x_0\}$.

Theorem 6.6.9. *Let* $f : I \times M \to M$ *be a mapping satisfying the following properties:*

(C.1) *for each* $x \in M$, *the mapping* $f(\cdot, x) \equiv [t \mapsto f(t, x)] : I \to M$ *is strongly measurable, and* $f(\cdot, y_0) \in L_\Phi(I; M)$ *for some* $y_0 \in M$;

(C.2) *there exists a constant* $L > 0$ *such that* $\|f(t, x) - f(t, y)\| \leq L\|x - y\|$ *for almost all* $t \in I$ *and all* $x, y \in M$.

Then, the integral operator T, *defined by*

$$(Tx)(t) = x_0 + \int_a^t f(s, x(s)) \, ds, \qquad x \in \dot{X}_w^*, \quad t \in I, \tag{6.6.10}$$

maps $\dot{X}_w^*$ *into itself, and the following inequality holds:*

$$w_{L|I|\lambda}(Tx, Ty) \leq w_\lambda(x, y) \quad \text{for all } \lambda > 0 \text{ and } x, y \in \dot{X}_w^*. \tag{6.6.11}$$

Proof. 1. First, we show that T is well-defined. Suppose $x \in \dot{X}_w^*$. Since $x \in$ $GV_\Phi(I; M)$ and $x(a) = x_0$, so that $x \in AC(I; M)$, conditions (C.1) and (C.2) imply that the composed mapping $t \mapsto f(t, x(t))$ is strongly measurable. Let us prove that this mapping belongs to $L_1(I; M)$. By Theorem 6.6.4, $x(t) = x_0 + \int_a^t x'(s)\, ds$ for all $t \in I$, and so, condition (C.2) yields

$$
\begin{aligned}
\|f(t, x(t))\| &\leq \|f(t, x(t)) - f(t, y_0)\| + \|f(t, y_0)\| \\
&\leq L\|x(t) - y_0\| + \|f(t, y_0)\| \\
&\leq L \int_I \|x'(s)\|\, ds + L\|x_0 - y_0\| + \|f(t, y_0)\| \qquad (6.6.12)
\end{aligned}
$$

for almost all $t \in I$. Since $x \in GV_\Phi(I; M)$, Theorem 6.6.4 implies the existence of a constant $\lambda_1 = \lambda_1(x) > 0$ such that

$$
C_1 \equiv V_{\Phi_{\lambda_1}}(x, I) = \int_I \Phi\left(\frac{\|x'(s)\|}{\lambda_1}\right) ds < \infty,
$$

and since, by (C.1), $f(\cdot, y_0) \in L_\Phi(I; M)$, there is $\lambda_2 = \lambda_2(f(\cdot, y_0)) > 0$ such that

$$
C_2 \equiv \rho(f(\cdot, y_0)/\lambda_2) = \int_I \Phi\left(\frac{\|f(t, y_0)\|}{\lambda_2}\right) dt < \infty.
$$

Setting $\lambda_0 = L\lambda_1|I| + 1 + \lambda_2$ and noting that

$$
\frac{L\lambda_1|I|}{\lambda_0} + \frac{1}{\lambda_0} + \frac{\lambda_2}{\lambda_0} = 1,
$$

by the convexity of Φ, we find (see (6.6.12))

$$
\Phi\left(\frac{1}{\lambda_0}\left[L \int_I \|x'(s)\|\, ds + L\|x_0 - y_0\| + \|f(t, y_0)\|\right]\right)
$$

$$
\leq \frac{L\lambda_1|I|}{\lambda_0} \Phi\left(\frac{1}{|I|} \int_I \frac{\|x'(s)\|}{\lambda_1}\, ds\right) + \frac{1}{\lambda_0} \Phi(L\|x_0 - y_0\|) + \frac{\lambda_2}{\lambda_0} \Phi\left(\frac{\|f(\cdot, y_0)\|}{\lambda_2}\right),
$$

and so, (6.6.12) and Jensen's integral inequality (A.1.3) yield

$$
\int_I \Phi\left(\frac{\|f(t, x(t))\|}{\lambda_0}\right) dt \leq \frac{L\lambda_1|I|}{\lambda_0} C_1 + \frac{|I|}{\lambda_0} \Phi(L\|x_0 - y_0\|) + \frac{\lambda_2}{\lambda_0} C_2, \qquad (6.6.13)
$$

where the right-hand side, denoted by C_0, is finite. Applying Jensen's integral inequality once again, we get

$$
\Phi\left(\frac{1}{\lambda_0|I|} \int_I \|f(t, x(t))\|\, dt\right) \leq \frac{1}{|I|} \int_I \Phi\left(\frac{\|f(t, x(t))\|}{\lambda_0}\right) dt \leq \frac{C_0}{|I|},
$$

which implies

$$\int_I \|f(t,x(t))\|\, dt \le \lambda_0 |I| \Phi^{-1}\left(\frac{C_0}{|I|}\right) < \infty.$$

Thus, $[t \mapsto f(t,x(t))] \in L_1(I;M)$. As a consequence, the operator T is well-defined on $\dot{X}_w^*$ and, by (6.6.10), $Tx \in AC(I;M)$ for all $x \in \dot{X}_w^*$, which implies that the almost everywhere derivative $(Tx)'$ belongs to $L_1(I;M)$ and satisfies the equality

$$(Tx)'(t) = f(t,x(t)) \quad \text{for almost all} \quad t \in I. \tag{6.6.14}$$

2. It is clear from (6.6.10) that, given $x \in \dot{X}_w^*$, we have $(Tx)(a) = x_0$. In order to show that $Tx \in GV_\Phi(I;M)$, we take into account (6.6.8), (6.6.14), and (6.6.13):

$$w_{\lambda_0}(Tx, 0) = \int_I \Phi\left(\frac{\|(Tx)'(t)\|}{\lambda_0}\right) dt = \int_I \Phi\left(\frac{\|f(t,x(t))\|}{\lambda_0}\right) dt \le C_0,$$

and so, T maps $\dot{X}_w^*$ into itself.
3. To prove inequality (6.6.11), let $\lambda > 0$ and $x,y \in \dot{X}_w^*$. By virtue of (6.6.8) and (6.6.14), we find

$$w_{L|I|\lambda}(Tx, Ty) = w_{L|I|\lambda}(Tx - Ty, 0) = \int_I \Phi\left(\frac{\|(Tx-Ty)'(t)\|}{L|I|\lambda}\right) dt$$

$$= \int_I \Phi\left(\frac{\|f(t,x(t)) - f(t,y(t))\|}{L|I|\lambda}\right) dt. \tag{6.6.15}$$

Since $x(a) = y(a) = x_0$, assumption (C.2) and Theorem 6.6.4 imply

$$\|f(t,x(t)) - f(t,y(t))\| \le L\|x(t) - y(t)\| \le L \int_I \|(x-y)'(s)\|\, ds,$$

for almost all $t \in I$, and so, Jensen's integral inequality, the monotonicity of Φ and (6.6.8) yield

$$\Phi\left(\frac{\|f(t,x(t)) - f(t,y(t))\|}{L|I|\lambda}\right) \le \Phi\left(\frac{1}{|I|}\int_I \frac{\|(x-y)'(s)\|}{\lambda}\, ds\right)$$

$$\le \frac{1}{|I|}\int_I \Phi\left(\frac{\|(x-y)'(s)\|}{\lambda}\right) ds = \frac{1}{|I|} w_\lambda(x,y).$$

Now, inequality (6.6.11) follows from (6.6.15). $\square$

Corollary 6.6.10. *Under the assumptions of Theorem 6.6.9, suppose $L|I| < 1$. Then, there is $x \in GV_\Phi(I; M)$ such that $x'(t) = f(t, x(t))$ for almost all $t \in I$ and $x(a) = x_0$.*

Proof. By Remark 6.6.6(5) and Theorem 6.1.6(c), the metric space $(\dot{X}_w^*, d_M)$ is complete, where $\dot{X}_w^* = \{x \in GV_\Phi(I; M) : x(a) = x_0\}$ and

$$d_M(x, y) = d(x(a), y(a)) + d_w^*(x, y) = d_w^*(x, y) \quad \text{for all} \quad x, y \in \dot{X}_w^*.$$

Let us show that the integral operator T from (6.6.10) is contractive on $\dot{X}_w^*$; more precisely, $d_w^*(Tx, Ty) \leq L|I| d_w^*(x, y)$ for all $x, y \in \dot{X}_w^*$. In fact, given $\lambda > 0$ such that $w_\lambda(x, y) \leq 1$, inequality (6.6.11) implies $w_{L|I|\lambda}(Tx, Ty) \leq 1$, and so, definition (2.3.3) of d_w^* gives $d_w^*(Tx, Ty) \leq L|I|\lambda$. Due to the arbitrariness of $\lambda > 0$ as above, the contractiveness of T follows. Applying the Banach Contraction Principle to the operator T on $\dot{X}_w^*$, we infer that T admits a fixed point, i.e., $Tx = x$ for some $x \in \dot{X}_w^*$. Finally, taking into account Theorem 6.6.4, we obtain the result. $\qquad \square$

6.7 Bibliographical Notes and Comments

Section 6.1. Functions of bounded φ-variation, generalizing functions of bounded variation in the sense of Jordan [49], were introduced by Wiener [104] and Young [107]. They were studied in detail by Musielak and Orlicz [76], Leśniewicz and Orlicz [62], and Ciemnoczołowski, Matuszewska and Orlicz [32]. Goffman, Moran and Waterman [42] showed that if $x \in Reg(I; \mathbb{R})$ and, at each point $t \in I$ of discontinuity of x, the value $x(t)$ lies between the values $\min\{x(t-0), x(t+0)\}$ and $\max\{x(t-0), x(t+0)\}$, then $V_\varphi(x, I) < \infty$ for some convex φ-function φ on $[0, \infty)$ such that $\varphi(u)/u \to 0$ as $u \to +0$. For a discussion and particular solutions of the problem from Remark 6.1.4 when $M = \mathbb{R}$, see Herda [45], Matuszewska and Orlicz [70], Wang and Wu [105]. Corollary 6.1.7 was proved by Musielak and Orlicz [76] (for $M = \mathbb{R}$), and Chistyakov [18]. In the exposition we follow Chistyakov [19, 25].

Section 6.2. Lipschitzian functions, mappings, and operators are classical in Functional Analysis. Lipschitzian operators basing on (6.2.1) were studied in Smajdor and Smajdor [99]. Our (modular) approach is from Chistyakov [25, Sect. 5.5].

Section 6.3. A treatise on the theory of superposition operators is Appell and Zabrejko [3]. In this section, we demonstrate the modular approach to superposition operators. Theorem 6.3.1 generalizes the Banach algebra property from Maligranda and Orlicz [69], and it is a particular case of the results in Chistyakov [19, 25].

Section 6.4. Continuous and Lipschitzian selections of multifunctions from an interval into a normed space with (nonconvex) compact values were established by Hermes [46]. Selections of bounded variation were discovered by Chistyakov [10]. Refinements of the selection theorem from [10] are presented in [7, 11, 12, 14]. A detailed study of selections of bounded (generalized) variation is Chistyakov [18].

Chistyakov and Galkin [30] showed that multifunctions of bounded φ-variation (even with $\varphi(u) = u^p$, $u \geq 0$, $p > 1$) may admit no selections of bounded φ-variation.

Section 6.5. This is an unusual modular approach to absolutely continuous functions and mappings, which are classically introduced by (6.5.4), e.g., Barbu and Precupanu [6], Natanson [83], Schwartz [97]. One more time it demonstrates the flexibility and applicability of metric modulars (even in the classical setting).

Section 6.6. The notion of the Φ-variation $V_\Phi(x, I)$ with $\Phi(u) = u^p$ ($p > 1$) and $M = \mathbb{R}$ was introduced by Riesz [94], and Medvedev [72] for general convex φ-functions Φ with $[\Phi] = \infty$, where they proved the equivalence as in Corollary 6.6.5(a). The integral formula (6.6.8) is due to Cybertowicz and Matuszewska [34] for $M = \mathbb{R}$, and Chistyakov [12, 14] for metric space (M, d). Further properties and generalizations of the Φ-variation are contained in Chistyakov [13, 15, 16]. Regular selections of classes AC and GV$_\Phi$ were established in Belov and Chistyakov [7] and Chistyakov [18]. The missing details in step 2 of the proof of Theorem 6.6.4 can be found in Barbu and Precupanu [6, Sect. 3.2]. The Banach algebra property of the space GV$_\Phi(I; \mathbb{R})$ was established by Maligranda and Orlicz [69]. Complete characterization of multivalued Lipschitzian superposition operators in classes GV$_\Phi$ is presented in Chistyakov [18, Sect. 13]. Theorem 6.6.3 and Corollary 6.6.5(b) are due to Chistyakov [14]. Examples 6.6.7 and 6.6.8 and Theorem 6.6.9 are taken from Chistyakov [28]. Theorem 6.6.9 supplements certain results in the theory of ordinary differential equations of Carathéodory's type (see Filippov [38]), where solutions x of class AC($I; M$) are usually considered under the assumption $f(\cdot, y_0) \in L_1(I; M)$.

Appendix

A.1 Superadditive, Subadditive, and Convex Functions

A function $\varphi : [0, \infty) \to [0, \infty)$ is said to be *superadditive* if

(a) $\varphi(+0) \equiv \lim_{u \to +0} \varphi(u) = 0$,
(b) $\varphi(u) > 0$ for $u > 0$, and
(c) $\varphi(u) + \varphi(v) \leq \varphi(u + v)$ for all $u, v \geq 0$.

Such a function φ is increasing, $\varphi(0) = 0$, and $\varphi(\infty) \equiv \lim_{u \to \infty} \varphi(u) = \infty$.
 The inverse function $\varphi^{-1} : [0, \infty) \to [0, \infty)$ of φ is well-defined, increasing, $\varphi^{-1}(\infty) = \infty$, and *subadditive*:

(a$_1$) $\varphi^{-1}(+0) = \varphi^{-1}(0) = 0$,
(b$_1$) $\varphi^{-1}(u) > 0$ for $u > 0$, and
(c$_1$) $\varphi^{-1}(u + v) \leq \varphi^{-1}(u) + \varphi^{-1}(v)$ for all $u, v \geq 0$.

Convex functions are an important particular case of superadditive functions. More precisely, if $\varphi : [0, \infty) \to [0, \infty)$ is a *convex* function such that $\varphi(u) = 0$ only at $u = 0$, then φ is *superadditive* and *continuous* on $[0, \infty)$, and the function $u \mapsto \varphi(u)/u$ is *nondecreasing* on $[0, \infty)$, so that

$$\lim_{u \to +0} \frac{\varphi(u)}{u} = \inf_{u>0} \frac{\varphi(u)}{u} \in [0, \infty) \quad \text{and} \quad \lim_{u \to \infty} \frac{\varphi(u)}{u} = \sup_{u>0} \frac{\varphi(u)}{u} \in (0, \infty].$$

In fact, the convexity of φ on $[0, \infty)$ means that, for all $0 \leq \alpha \leq 1$ and $u, v \geq 0$,

$$\varphi(\alpha u + (1 - \alpha)v) \leq \alpha\varphi(u) + (1 - \alpha)\varphi(v), \qquad \text{(A.1.1)}$$

© Springer International Publishing Switzerland 2015
V.V. Chistyakov, *Metric Modular Spaces*, SpringerBriefs in Mathematics,
DOI 10.1007/978-3-319-25283-4

and, in particular $\varphi(\alpha u) \leq \alpha\varphi(u)$. Given $0 \leq u_1 < u_2$, setting $\alpha = u_1/u_2$ and $u = u_2$, we get $\varphi(u_1)/u_1 \leq \varphi(u_2)/u_2$. Since

$$\varphi(u_i) \leq \frac{u_i}{u_1 + u_2} \varphi(u_1 + u_2) \quad \text{for} \ \ u_i \geq 0 \ \ \text{with} \ \ i = 1, 2,$$

the superadditivity property of φ follows.

In order to see that φ is continuous on $[0, \infty)$, we first note that $0 \leq \varphi(\alpha) \leq \alpha\varphi(1)$ for $0 \leq \alpha \leq 1$, and so, $\varphi(+0) = 0 = \varphi(0)$. Now, suppose $u > 0$. Since φ is increasing, the one-sided limits $\varphi(u - 0)$ and $\varphi(u + 0)$ exist in $[0, \infty)$, and we have $\varphi(u - 0) \leq \varphi(u) \leq \varphi(u + 0)$. From the convexity of φ, we get

$$\frac{\varphi(u) - \varphi(u_1)}{u - u_1} \leq \frac{\varphi(u_2) - \varphi(u)}{u_2 - u} \quad \text{for all} \ \ 0 < u_1 < u < u_2,$$

that is,

$$\varphi(u) \leq \varphi(u_1) + (u - u_1) \frac{\varphi(u_2) - \varphi(u)}{u_2 - u} .$$

Passing to the limit as $u_1 \to u - 0$, we find $\varphi(u) \leq \varphi(u - 0)$. Similarly,

$$\frac{\varphi(u_2) - \varphi(u)}{u_2 - u} \leq \frac{\varphi(u_3) - \varphi(u_2)}{u_3 - u_2} \quad \text{if} \ \ u < u_2 < u_3,$$

which implies

$$\varphi(u_2) - (u_2 - u) \frac{\varphi(u_3) - \varphi(u_2)}{u_3 - u_2} \leq \varphi(u).$$

Letting u_2 go to $u + 0$, we get $\varphi(u + 0) \leq \varphi(u)$, which completes the proof of the continuity of φ on $[0, \infty)$.

For instance, given $u, v \geq 0$, we have:

$$u^p + v^p \leq (u + v)^p \leq 2^{p-1}(u^p + v^p) \quad \text{if} \ \ p \geq 1,$$

$$(u + v)^p \leq u^p + v^p \leq 2^{1-p}(u + v)^p \quad \text{if} \ \ 0 < p < 1.$$

Let $\varphi : [0, \infty) \to [0, \infty)$ be a *convex* function such that $\varphi(u) = 0$ iff $u = 0$. Its inverse φ^{-1} is increasing, continuous, and *concave* on $[0, \infty)$. In fact, given $u_1, v_1 \geq 0$, we set $u = \varphi^{-1}(u_1)$ and $v = \varphi^{-1}(v_1)$, so that $u_1 = \varphi(u)$ and $v_1 = \varphi(v)$. By (A.1.1),

$$\varphi\big(\alpha\varphi^{-1}(u_1) + (1-\alpha)\varphi^{-1}(v_1)\big) \leq \alpha u_1 + (1-\alpha)v_1, \qquad 0 \leq \alpha \leq 1,$$

whence, taking the inverse function φ^{-1} from both sides of this inequality,

$$\alpha\varphi^{-1}(u_1) + (1-\alpha)\varphi^{-1}(v_1) \leq \varphi^{-1}(\alpha u_1 + (1-\alpha)v_1).$$

Note that since φ^{-1} is concave, it is subadditive.

The function $\omega_\varphi(u) = u\varphi^{-1}(1/u)$, $u > 0$, is continuous, *nondecreasing*, and *concave* on $(0, \infty)$. In fact, given $0 < u_1 < u_2$, we set $v_1 = \varphi^{-1}(1/u_1)$ and $v_2 = \varphi^{-1}(1/u_2)$. Since $1/u_2 < 1/u_1$ and φ^{-1} is increasing, we get $v_2 < v_1$, and since the function $v \mapsto \varphi(v)/v$ is nondecreasing, we find $\varphi(v_2)/v_2 \le \varphi(v_1)/v_1$ or, equivalently, $v_1/\varphi(v_1) \le v_2/\varphi(v_2)$. Noting that $1/\varphi(v_1) = u_1$ and $1/\varphi(v_2) = u_2$, we have

$$\omega_\varphi(u_1) = u_1\varphi^{-1}(1/u_1) \le u_2\varphi^{-1}(1/u_2) = \omega_\varphi(u_2).$$

The concavity of ω_φ on $(0, \infty)$ means that

$$\omega_\varphi(\alpha u + (1-\alpha)v) \ge \alpha\omega_\varphi(u) + (1-\alpha)\omega_\varphi(v), \qquad \alpha \in [0,1], \quad u, v > 0,$$

or, more explicitly,

$$(\alpha u + (1-\alpha)v)\varphi^{-1}\left(\frac{1}{\alpha u + (1-\alpha)v}\right) \ge \alpha u\varphi^{-1}\left(\frac{1}{u}\right) + (1-\alpha)v\varphi^{-1}\left(\frac{1}{v}\right),$$

which is equivalent to

$$\varphi^{-1}\left(\frac{1}{\alpha u + (1-\alpha)v}\right) \ge \frac{\alpha u}{\alpha u + (1-\alpha)v}\,\varphi^{-1}\left(\frac{1}{u}\right) + \frac{(1-\alpha)v}{\alpha u + (1-\alpha)v}\,\varphi^{-1}\left(\frac{1}{v}\right).$$

The last inequality follows from the concavity of function φ^{-1}, because

$$\frac{1}{\alpha u + (1-\alpha)v} = \frac{\alpha u}{\alpha u + (1-\alpha)v}\cdot\frac{1}{u} + \frac{(1-\alpha)v}{\alpha u + (1-\alpha)v}\cdot\frac{1}{v}.$$

It is to be noted that

$$\omega_\varphi(+0) = \lim_{u\to+0}\omega_\varphi(u) = \lim_{u\to+0}u\varphi^{-1}\left(\frac{1}{u}\right) = \lim_{v\to\infty}\frac{v}{\varphi(v)} \in [0, \infty).$$

Since φ is convex, the following *Jensen's inequalities* hold:

$$\varphi\left(\frac{\sum_{i=1}^n\alpha_i u_i}{\sum_{i=1}^n\alpha_i}\right) \le \frac{\sum_{i=1}^n\alpha_i\varphi(u_i)}{\sum_{i=1}^n\alpha_i} \tag{A.1.2}$$

where $\alpha_i \ge 0$, $u_i \ge 0$, $i = 1, \ldots, n$, are such that $\sum_{i=1}^n\alpha_i > 0$, and

$$\varphi\left(\frac{1}{b-a}\int_a^b |f(t)|\,dt\right) \le \frac{1}{b-a}\int_a^b \varphi(|f(t)|)\,dt, \tag{A.1.3}$$

where $f : [a, b] \to [-\infty, \infty]$ is a measurable almost everywhere finite function such that the Lebesgue integrals make sense.

For more information, see Krasnosel'skiĭ and Rutickiĭ [56], Natanson [83], and Schwartz [97].

A.2 The Hausdorff Distance

Let (X, d) be a metric space. If $\varnothing \neq A, B \subset X$, the *excess of A over B* is the quantity

$$e(A, B) = \sup_{x \in A} \text{dist}(x, B) \in [0, \infty], \qquad (A.2.1)$$

where $\text{dist}(x, B) = \inf_{y \in B} d(x, y)$ is the distance from the point x to the set B (and $\text{dist}(x, \varnothing) = \infty$). We also set $e(\varnothing, B) = 0$ for all $B \subset X$, and $e(A, \varnothing) = \infty$ for all $\varnothing \neq A \subset X$. Alternatively, the excess $e(A, B)$ can be expressed as

$$e(A, B) = \inf\{r > 0 : A \subset \mathcal{O}_r(B)\},$$

where $\mathcal{O}_r(B) = \{x \in X : \text{dist}(x, B) < r\}$ is the *r-neighborhood* of the set B, i.e., the union of all open balls in X with centers at points of B and the same radius r (and $\mathcal{O}_r(\varnothing) = \varnothing$ for any $r > 0$). The excess has two main properties: given $A, B, C \subset X$,

(a) $e(A, B) = 0$ iff $A \subset \overline{B}$, where $\overline{B} = \bigcap_{r > 0} \mathcal{O}_r(B)$ is the closure of B in X;
(b) $e(A, C) \leq e(A, B) + e(B, C)$.

The *Hausdorff distance* between any $A \subset X$ and $B \subset X$ is defined by

$$D(A, B) \equiv D_d(A, B) = \max\{e(A, B), e(B, A)\}$$
$$= \inf\{r > 0 : A \subset \mathcal{O}_r(B) \text{ and } B \subset \mathcal{O}_r(A)\},$$

and the following properties hold, for any $A, B, C \subset X$:

(A) $D(A, B) = 0$ iff $\overline{A} = \overline{B}$;
(B) $D(A, B) = D(B, A)$;
(C) $D(A, C) \leq D(A, B) + D(B, C)$.

Note also that $D(\varnothing, \varnothing) = 0$, and $D(A, \varnothing) = \infty$ for all $\varnothing \neq A \subset X$. It follows that:

- D is an extended pseudometric on $\mathscr{P}(X)$, the family of all subsets of X;
- D is an extended metric on $\text{cl}(X)$, the family of all *closed* subsets of X (with or without the empty set);
- D is a metric on $\text{cb}(X)$, the family of all *nonempty closed bounded* subsets of X, called the *Hausdorff metric* (induced by d), and in particular,
- D is a metric on $\text{c}(X)$, the family of all *nonempty compact* subsets of X.

The construction of the Hausdorff distance remains valid if d is an extended (pseudo)metric on X.

For more information, see Castaing and Valadier [9], Hausdorff [43], and Kuratowski [58].

A.3   Metric Semigroups and Abstract Convex Cones

1. A triple $(M, d, +)$ is said to be a *metric semigroup* if (M, d) is a metric space with metric d, $(M, +)$ is an Abelian semigroup with the operation of addition $+$, and d is translation invariant with respect to $+$ in the sense that $d(x + z, y + z) = d(x, y)$ for all $x, y, z \in M$. An element $0 \in M$ such that $x + 0 = 0 + x = x$ for all $x \in M$ is called the *zero* in M (it is determined uniquely).

Given a metric semigroup $(M, d, +)$ and elements $x, y, \bar{x}, \bar{y} \in M$, we have:

$$d(x, y) \leq d(x + \bar{x}, y + \bar{y}) + d(\bar{x}, \bar{y}),$$

$$d(x + \bar{x}, y + \bar{y}) \leq d(x, y) + d(\bar{x}, \bar{y}). \tag{A.3.1}$$

If sequences $\{x_n\}$, $\{y_n\}$, $\{\bar{x}_n\}$, and $\{\bar{y}_n\}$ of elements from M converge in M to elements $x, y, \bar{x}$, and $\bar{y}$ as $n \to \infty$, respectively, then, by virtue of (A.3.1),

$$\lim_{n \to \infty} d(x_n + \bar{x}_n, y_n + \bar{y}_n) = d(x + \bar{x}, y + \bar{y}),$$

and, in particular, the addition operation $(x, y) \mapsto x + y$ is a continuous mapping from $M \times M$ into M.

Particular cases of metric semigroups are metric linear spaces in Rolewicz [95].

2. A quadruple $(M, d, +, \cdot)$ is said to be an *abstract convex cone* if $(M, d, +)$ is a metric semigroup with zero $0 \in M$, and the operation $\cdot : [0, \infty) \times M \to M$ of multiplication of numbers $\alpha \geq 0$ by elements $x \in M$, written as $(\alpha, x) \mapsto \alpha \cdot x \equiv \alpha x$, has the following properties, for all $\alpha, \beta \geq 0$ and $x, y \in M$:

$$d(\alpha x, \alpha y) = \alpha d(x, y),$$

$$\alpha(x + y) = \alpha x + \alpha y, \quad (\alpha + \beta)x = \alpha x + \beta x, \quad \alpha(\beta x) = (\alpha \beta)x, \quad 1 \cdot x = x$$

(see Chistyakov [18–20], Smajdor [100]).

A metric semigroup $(M, d, +)$ or an abstract convex cone $(M, d, +, \cdot)$ is said to be *complete* if the underlying metric space (M, d) is complete.

Given an abstract convex cone $(M, d, +, \cdot)$, $\alpha, \beta \geq 0$, and $x, y \in M$, we have

$$d(\alpha x + \beta y, \beta x + \alpha y) = |\alpha - \beta| d(x, y), \tag{A.3.2}$$

which implies

$$d(\alpha x, \beta y) \leq \alpha d(x, y) + |\alpha - \beta| d(y, 0), \tag{A.3.3}$$

and so, the operation of multiplication $(\alpha, x) \mapsto \alpha x$ is a continuous mapping from $[0, \infty) \times M$ into M.

Note that $\alpha \cdot 0 = 0$ in M for all $\alpha \geq 0$: in fact, by (A.3.3),

$$d(\alpha \cdot 0, 0) = d(\alpha \cdot 0, 1 \cdot 0) \leq \alpha d(0, 0) + |\alpha - 1| d(0, 0) = 0.$$

Denote by $|x| \equiv |x|_d = d(x, \mathbf{0})$ the (so called) 'absolute value' of $x \in M$. If $x \in M$, $x \neq \mathbf{0}$, and $x' = (1/|x|) \cdot x \equiv x/|x|$, then $|x'| = 1$:

$$|x'| = d(x', \mathbf{0}) = d\left(\frac{x}{|x|}, \frac{\mathbf{0}}{|x|}\right) = \frac{1}{|x|} d(x, \mathbf{0}) = \frac{1}{|x|} \cdot |x| = 1.$$

A simple example of an abstract convex cone is a normed linear space $(X, \|\cdot\|)$ with the induced metric $d(x, y) = \|x - y\|$, $x, y \in X$, and operations $+$ and $\cdot$ from X. If $K \subset X$ is a convex cone (i.e., $x + y$, $\alpha x \in K$ for all $x, y \in K$ and $\alpha \geq 0$), then $(K, d, +, \cdot)$ is an abstract convex cone, which is complete if X is a Banach space and K is closed in X.

One more example is as follows. Suppose $(X, \|\cdot\|)$ is a real linear space. Denote by cbc (X) the family of all nonempty closed bounded convex subsets of X. Given $A, B \in$ cbc (X), we set $A + B = \{x + y : x \in A, y \in B\}$ (Minkowski's sum), $\alpha A = \{\alpha x : x \in A\}$ if $\alpha \geq 0$, and $A \boxplus B = \text{cl}(A + B)$, where cl (C) designates the closure in X of the set $C \subset X$. The introduced operations in cbc (X) have the properties (see Hörmander [47], Pinsker [86]): $A \boxplus B = \text{cl}(\text{cl}(A) + \text{cl}(B))$, $\alpha(A \boxplus B) = (\alpha A) \boxplus (\alpha B)$, $(\alpha + \beta)A = (\alpha A) \boxplus (\beta A)$, $\alpha(\beta A) = (\alpha \beta)A$, and $1 \cdot A = A$ for all $\alpha, \beta \geq 0$. The Abelian semigroup cbc (X) is endowed with the Hausdorff metric D, generated by the norm $\|\cdot\|$ in X, and so, $D(A, B)$ can be written as

$$D(A, B) = \inf \{\alpha > 0 : A \subset B + \alpha S \text{ and } B \subset A + \alpha S\},$$

where $S = \{x \in X : \|x\| \leq 1\}$ is the unit ball in X. Additional properties of D are as follows (see De Blasi [35], Rådström [91]): if $A, B, C \in$ cbc (X) and $\alpha \geq 0$, then

$$D(\alpha A, \alpha B) = \alpha D(A, B),$$

$$D(A \boxplus C, B \boxplus C) = D(A + C, B + C) = D(A, B).$$

Consequently, the quadruple (cbc $(X), D, \boxplus, \cdot)$ is an abstract convex cone, which is complete provided X is a Banach space (see Castaing and Valadier [9]).

More examples of abstract convex cones are presented in Chistyakov [17–19, 24, 25].

References

1. Adams, R.A.: Sobolev Spaces. Pure and Applied Mathematics, vol. 65. Academic, New York (1975)
2. Aleksandrov, P.S.: Introduction to Set Theory and General Topology. Nauka, Moscow (1977) (in Russian) German translation: VEB Deutscher Verlag der Wissenschaften, Berlin (1984)
3. Appell, J., Zabrejko, P.P.: Nonlinear Superposition Operators. Cambridge University Press, Cambridge (1990)
4. Banach, S.: Sur les opérations dans les ensembles abstraits et leur application aux équations intégrales. Fundam. Math. **3**, 133–181 (1922) (in French)
5. Banach, S.: Théorie des Opérations Linéaires. Monografie Matematyczne 1, Warszawa (1932) (in French)
6. Barbu, V., Precupanu, Th.: Convexity and Optimization in Banach Spaces. Revised edition. Sijthoff & Noordhoff International Publishers, Alphen aan den Rijn (1978)
7. Belov, S.A., Chistyakov, V.V.: A selection principle for mappings of bounded variation. J. Math. Anal. Appl. **249**(2), 351–366 (2000)
8. Bielecki, A.: Une remarque sur la méthode de Banach-Cacciopoli-Tikhonov dans la théorie des équations différentielles ordinaires. Bull. Acad. Polon. Sci. Cl. III. **4**, 261–264 (1956) (in French)
9. Castaing, C., Valadier, M.: Convex Analysis and Measurable Multifunctions. Lecture Notes in Mathematics, vol. 580. Springer, Berlin (1977)
10. Chistyakov, V.V.: On mappings of bounded variation. J. Dyn. Control Syst. **3**(2), 261–289 (1997)
11. Chistyakov, V.V.: On the theory of multivalued mappings of bounded variation of one real variable. Mat. Sb. **189**(5), 153–176 (1998) (in Russian) English translation: Sbornik Math. **189**(5–6), 797–819 (1998)
12. Chistyakov, V.V.: Mappings of bounded variation with values in a metric space: generalizations. Pontryagin Conference, 2, Nonsmooth Analysis and Optimization (Moscow, 1998). J. Math. Sci. (N.Y.) **100**(6), 2700–2715 (2000)
13. Chistyakov, V.V.: Lipschitzian superposition operators between spaces of functions of bounded generalized variation with weight. J. Appl. Anal. **6**(2), 173–186 (2000)
14. Chistyakov, V.V.: Generalized variation of mappings with applications to composition operators and multifunctions. Positivity **5**(4), 323–358 (2001)
15. Chistyakov, V.V.: Metric space-valued mappings of bounded variation. Functional analysis, 8. J. Math. Sci. (N.Y.) **111**(2), 3387–3429 (2002)
16. Chistyakov, V.V.: On multi-valued mappings of finite generalized variation. Mat. Zametki **71**(4), 611–632 (2002) (in Russian) English translation: Math. Notes **71**(3–4), 556–575 (2002)

© Springer International Publishing Switzerland 2015

V.V. Chistyakov, *Metric Modular Spaces*, SpringerBriefs in Mathematics,

DOI 10.1007/978-3-319-25283-4

17. Chistyakov, V.V.: Metric semigroups and cones of mappings of finite variation of several variables, and multivalued superposition operators. Dokl. Akad. Nauk **393**(6), 757–761 (2003) (in Russian) English translation: Dokl. Math. **68**(3), 445–448 (2003)

18. Chistyakov, V.V.: Selections of bounded variation. J. Appl. Anal. **10**(1), 1–82 (2004)

19. Chistyakov, V.V.: Lipschitzian Nemytskii operators in the cones of mappings of bounded Wiener φ-variation. Folia Math. **11**(1), 15–39 (2004)

20. Chistyakov, V.V.: Abstract superposition operators on mappings of bounded variation of two real variables. I. Sibirsk. Mat. Zh. **46**(3), 698–717 (2005) (in Russian) English traslation: Siberian Math. J. **46**(3), 555–571 (2005)

21. Chistyakov, V.V.: The optimal form of selection principles for functions of a real variable. J. Math. Anal. Appl. **310**(2), 609–625 (2005)

22. Chistyakov, V.V.: Metric modulars and their application. Dokl. Akad. Nauk **406**(2), 165–168 (2006) (in Russian) English translation: Dokl. Math. **73**(1), 32–35 (2006)

23. Chistyakov, V.V.: Modular metric spaces generated by F-modulars. Folia Math. **15**(1), 3–24 (2008)

24. Chistyakov, V.V.: Modular metric spaces, I: Basic concepts. Nonlinear Anal. **72**(1), 1–14 (2010)

25. Chistyakov, V.V.: Modular metric spaces, II: Application to superposition operators. Nonlinear Anal. **72**(1), 15–30 (2010)

26. Chistyakov, V.V.: A fixed point theorem for contractions in modular metric spaces. e-Print. arXiv: 1112.5561, 1–31 (2011)

27. Chistyakov, V.V.: Fixed points of modular contractive maps. Dokl. Akad. Nauk **445**(3), 274–277 (2012) (in Russian) English translation: Dokl. Math. **86**(1), 515–518 (2012)

28. Chistyakov, V.V.: Modular contractions and their application. In: Models, Algorithms, and Technologies for Network Analysis. Springer Proceedings in Mathematics & Statistics, vol. 32, pp. 65–92. Springer, New York (2013)

29. Chistyakov, V.V.: Modular Lipschitzian and contractive maps. In: Migdalas, A., Karakitsiou, A. (eds.) Optimization, Control, and Applications in the Information Age. Springer Proceedings in Mathematics & Statistics, vol. 130, pp. 1–15. Springer International Publishing Switzerland (2015)

30. Chistyakov, V.V., Galkin, O.E.: On maps of bounded p-variation with $p > 1$. Positivity **2**(1), 19–45 (1998)

31. Chistyakov, V.V., Maniscalco, C., Tretyachenko, Yu.V.: Variants of a selection principle for sequences of regulated and non-regulated functions. In: De Carli, L., Kazarian, K., Milman, M. (eds.) Topics in Classical Analysis and Applications in Honor of Daniel Waterman, pp. 45–72. World Scientific Publishing, Hackensack (2008)

32. Ciemnoczołowski, J., Matuszewska, W., Orlicz, W.: Some properties of functions of bounded φ-variation and of bounded φ-variation in the sense of Wiener. Bull. Pol. Acad. Sci. Math. **35**(3–4), 185–194 (1987)

33. Copson, E.T.: Metric Spaces. Cambridge Tracts in Mathematics and Mathematical Physics, No. 57. Cambridge University Press, London (1968)

34. Cybertowicz, Z., Matuszewska, W.: Functions of bounded generalized variations. Comment. Math. Prace Mat. **20**, 29–52 (1977)

35. De Blasi, F.S.: On the differentiability of multifunctions. Pac. J. Math. **66**(1), 67–81 (1976)

36. Deza, M.M., Deza, E.: Encyclopedia of Distances, 2nd edn. Springer, Berlin (2013)

37. Dudley R.M., Norvaiša, R.: Differentiability of Six Operators on Nonsmooth Functions and p-Variation. Springer, Berlin (1999)

38. Filippov, A.F.: Differential Equations with Discontinuous Right-Hand Sides. Nauka, Moscow (1985) (in Russian) English translation: Mathematics and Applications, vol. 18. Kluwer, Dordrecht (1988)

39. Fréchet, M.: Sur quelques points du calcul functionnel. Rend. Circ. Mat. Palermo **22**(1), 1–72 (1906) (in French)

40. Gniłka, S.: Modular spaces of functions of bounded M-variation. Funct. Approx. Comment. Math. **6**, 3–24 (1978)

41. Goebel, K., Kirk, W.A.: Topics in Metric Fixed Point Theory. Cambridge Studies in Advanced Mathematics, vol. 28. Cambridge University Press, Cambridge (1990)
42. Goffman, C., Moran, G., Waterman, D.: The structure of regulated functions. Proc. Am. Math. Soc. **57**(1), 61–65 (1976)
43. Hausdorff, F.: Grundzüge der Mengenlehre. Veit and Company, Leipzig (1914) (in German) English translation: Set Theory, 3rd edn. Chelsea Publishing, AMS, Providence (2005)
44. Helly, E.: Über lineare Funktionaloperationen. Kl. Kaiserlichen Akad. Wiss. Wien **121**, 265–297 (1912) (in German)
45. Herda, H.-H.: Modular spaces of generalized variation. Studia Math. **30**, 21–42 (1968)
46. Hermes, H.: On continuous and measurable selections and the existence of solutions of generalized differential equations. Proc. Am. Math. Soc. **29**(3), 535–542 (1971)
47. Hörmander, L.: Sur la fonction d'appui des ensembles convexes dans un espace localement convexe. Ark. Math. **3**(12), 181–186 (1954) (in French)
48. Hudzik, H., Maligranda, L.: Amemiya norm equals Orlicz norm in general. Indug. Mathem., N.S. **11**(4), 573–585 (2000)
49. Jordan, C.: Sur la série de Fourier. C. R. Acad. Sci. **92**(5), 228–230 (1881) Reprinted in Oeuvres, Gauthier-Villars **4**, 393–395 (1964) (in French)
50. Jung, C.F.K.: On generalized complete metric spaces. Bull. Am. Math. Soc. **75**, 113–116 (1969)
51. Kaplansky, I.: Set Theory and Metric Spaces. Allyn and Bacon, Boston (1972)
52. Kelley, J.L.: General Topology (reprint of the 1955 Edition, Van Nostrand, Toronto). Graduate Texts in Mathematics, No. 27. Springer, New York (1975)
53. Koshi, S., Shimogaki, T.: On F-norms of quasi-modular spaces. J. Fac. Sci. Hokkaido Univ. Ser. I **15**(3–4), 202–218 (1961)
54. Kolmogorov, A.N., Fomin, S.V.: Elements of the Theory of Functions and Functional Analysis, 6th edn. Nauka, Moscow (1989) (in Russian) English translation of the First Edition: Metric and Normed Spaces, vol. 1. Graylock Press, Rochester (1957)
55. Kozlowski, W.M.: Modular Function Spaces. Monographs and Textbooks in Pure and Applied Mathematics, vol. 122. Marcel Dekker, New York (1988)
56. Krasnosel'skiĭ, M.A., Rutickiĭ, Ja.B.: Convex Functions and Orlicz Spaces. Fizmatgiz, Moscow (1958) (in Russian) English translation: P. Noordhoff Ltd., Groningen (1961)
57. Kumaresan, S.: Topology of Metric Spaces. Narosa Publishing House, New Delhi (2005)
58. Kuratowski, K.: Topology, vol. I. Academic, New York (1966)
59. Lebesgue, H.: Integrále, longueur, aire. Ann. Mat. Pura Appl. **7**, 231–259 (1902)
60. Leśniewicz, R.: On generalized modular spaces. I. Comment. Math. Prace Mat. **18**(2), 223–242 (1974/1975)
61. Leśniewicz, R.: On generalized modular spaces. II. Comment. Math. Prace Mat. **18**(2), 243–271 (1974/75)
62. Leśniewicz, R., Orlicz, W.: On generalized variations. II. Studia Math. **45**, 71–109 (1973) Reprinted in [89]: pp. 1434–1472
63. Leśniewicz, R., Orlicz, W.: A note on modular spaces. XIV. Bull. Acad. Polon. Sci. Sér. Sci. Math. Astronom. Phys. **22**, 915–923 (1974) Reprinted in [89]: pp. 1479–1487
64. Lindenbaum, A.: Contributions à l'étude de l'espace métrique. Fundam. Math. **8**, 209–222 (1926) (in French)
65. Lindenstrauss, J., Tzafriri, L.: Classical Banach Spaces. II. Function Spaces. Springer, Berlin (1979)
66. Luxemburg, W.A.J.: Banach function spaces. Thesis, Delft Technical University (1955)
67. Luxemburg, W.A.J.: On the convergence of successive approximations in the theory of ordinary differential equations. II. Nederl. Akad. Wetensch. Proc. Ser. A 61 = Indag. Math. **20**, 540–546 (1958)
68. Maligranda, L.: Orlicz Spaces and Interpolation. Seminars in Mathematics, vol. 5. Universidade Estadual de Campinas, Campinas SP (1989)
69. Maligranda, L., Orlicz, W.: On some properties of functions of generalized variation. Monatsh. Math. **104**, 53–65 (1987)

70. Matuszewska, W., Orlicz, W.: On property B_1 for functions of bounded φ-variation. Bull. Pol. Acad. Sci. Math. **35**(1–2), 57–69 (1987)

71. Mazur, S., Orlicz, W.: On some classes of linear spaces. Studia Math. **17**, 97–119 (1958) Reprinted in [89]: pp. 981–1003

72. Medvedev, Yu.T.: Generalization of a theorem of F. Riesz. Uspekhi Mat. Nauk **8**(6), 115–118 (1953) (in Russian)

73. Morse, M., Transue, W.: Functionals F bilinear over the product $A \times B$ of two pseudo-normed vector spaces. II. Admissible spaces A. Ann. Math. (2) **51**(3), 576–614 (1950)

74. Musielak, J.: A generalization of F-modular spaces. Beitr. Anal. **6**, 49–53 (1974)

75. Musielak, J.: Orlicz Spaces and Modular Spaces. Lecture Notes in Mathematics, vol. 1034. Springer, Berlin (1983)

76. Musielak, J., Orlicz, W.: On generalized variations (I). Studia Math. **18**, 11–41 (1959) Reprinted in [89]: pp. 1021–1051

77. Musielak, J., Orlicz, W.: On modular spaces. Studia Math. **18**, 49–65 (1959) Reprinted in [89]: pp. 1052–1068

78. Musielak, J., Orlicz, W.: Some remarks on modular spaces. Bull. Acad. Polon. Sci. Sér. Sci. Math. Astron. Phys. **7**, 661–668 (1959) Reprinted in [89]: pp. 1099–1106

79. Musielak, J., Peetre, J.: F-modular spaces. Funct. Approx. Comment. Math. **1**, 67–73 (1974)

80. Nakano, H.: Modulared Semi-Ordered Linear Spaces. Maruzen, Tokyo (1950)

81. Nakano, H.: Topology of Linear Topological Spaces. Maruzen, Tokyo (1951)

82. Nakano, H.: Generalized modular spaces. Studia Math. **31**, 439–449 (1968)

83. Natanson, I.P.: Theory of Functions of a Real Variable, 3rd edn. Nauka, Moscow (1974) (in Russian) English translation: Frederick Ungar Publishing Co., New York (1965)

84. Nowak, M.: Orlicz lattices with modular topology. I. Comment. Math. Univ. Carolin. **30**(2), 261–270 (1989)

85. Nowak, M.: Orlicz lattices with modular topology. II. Comment. Math. Univ. Carolin. **30**(2), 271–279 (1989)

86. Pinsker, A.G.: The space of convex sets of a locally convex space. In: Collection of papers of Leningrad. engineer.-econom. inst. named after P. Togliatti, vol. 63, pp. 3–17 (1966) (in Russian)

87. Orlicz, W.: Über eine gewisse Klasse von Räumen vom Typus B. Bull. Int. Acad. Polon. Sci. Sér. A, 207–220 (1932) Reprinted in [89]: pp. 217–230 (in German)

88. Orlicz, W.: Über Räume (L^M). Bull. Int. Acad. Polon. Sci. Sér. A, 93–107 (1936) Reprinted in [89]: pp. 345–359 (in German)

89. Orlicz, W.: Collected Papers. Parts I, II. PWN—Polish Scientific Publishers, Warsaw (1988)

90. Orlicz, W.: A note on modular spaces. I. Bull. Acad. Polon. Sci. Sér. Sci. Math. Astron. Phys. **9**, 157–162 (1961) Reprinted in [89]: pp. 1142–1147

91. Rådström, H.: An embedding theorem for spaces of convex sets. Proc. Am. Math. Soc. **3**(1), 165–169 (1952)

92. Rao, M.M., Ren, Z.D.: Theory of Orlicz Spaces. Monographs and Textbooks in Pure and Applied Mathematics, vol. 146. Marcel Dekker, New York (1991)

93. Rao, M.M., Ren, Z.D.: Applications of Orlicz Spaces. Monographs and Textbooks in Pure and Applied Mathematics, vol. 250. Marcel Dekker, New York (2002)

94. Riesz, F.: Untersuchungen über Systeme integrierbarer Funktionen. Ann. Math. **69**, 449–497 (1910) (in German)

95. Rolewicz, S.: Metric Linear Spaces. PWN, Reidel, Dordrecht, Warszawa (1985)

96. Schramm, M.: Functions of Φ-bounded variation and Riemann-Stieltjes integration. Trans. Am. Math. Soc. **287**(1), 49–63 (1985)

97. Schwartz, L.: Analyse Mathématique. Hermann, Paris (1967) (in French)

98. Shirali, S., Vasudeva, H.L.: Metric Spaces. Springer, London (2006)

99. Smajdor, A., Smajdor, W.: Jensen equation and Nemytski operator for set-valued functions. Rad. Mat. **5**, 311–320 (1989)

100. Smajdor, W.: Note on Jensen and Pexider functional equations. Demonstratio Math. **32**(2), 363–376 (1999)

101. Tret'yachenko, Yu.V., Chistyakov, V.V.: The selection principle for pointwise bounded sequences of functions. Mat. Zametki **84**(3), 428–439 (2008) (in Russian) English translation: Math. Notes **84**(4), 396–406 (2008)
102. Turpin, Ph.: Fubini inequalities and bounded multiplier property in generalized modular spaces. Special issue dedicated to Władysław Orlicz on the occasion of his seventy-fifth birthday. Comment. Math. Special Issue **1**, 331–353 (1978)
103. Waterman, D.: On Λ-bounded variation. Studia Math. **57**(1), 33–45 (1976)
104. Wiener, N.: The quadratic variation of a function and its Fourier coefficients. Massachusetts J. Math. Phys. **3**, 72–94 (1924)
105. Wang, J., Wu, C.: On a property of ϕ-variational modular spaces. Opusc. Math. **30**(2), 209–215 (2010)
106. Yamamuro, S.: On conjugate spaces of Nakano spaces. Trans. Am. Math. Soc. **90**, 291–311 (1959)
107. Young, L.C.: Sur une généralisation de la notion de variation p-ième bornée au sens de N. Wiener, et sur la convergence des séries de Fourier. C. R. Acad. Sci. Paris **204**(7), 470–472 (1937) (in French)

Index

Symbols
F-modular, 49
F-norm, 21, 30
Δ_2-condition, 71
φ-function, 93
φ-variation, 95
g-scaling of w, 49
g-truncation of w, 49

A
abstract convex cone, 47, 127
additivity, 103
almost everywhere, 114

B
ball
    closed, 68
    open, 66
Banach's contraction principle, 103, 121

C
Carathéodory's equation, 118
Cardano's formula, 25
Cauchy's condition, 82
Cauchy's equation, 100
center, 19
closed with respect to the
    metric convergence, 65
    modular convergence, 70
closure, 68
convergence
    metric, 65
    modular, 69

pointwise, 82
uniform, 82
convex right inverse, 60

D
diameter, 79

E
equivalence, 19
    nonlinear, 30
excess, 126

F
function
    concave, 50, 124
    convex, 123
    subadditive, 123
    superadditive, 123

G
generator, 100

H
Hausdorff distance, 126
Helly's selection theorem, 87

I
increment, 79
    joint, 79
interior, 67

J
Jensen's inequality, 125

L
left inverse of
   nondecreasing function Φ, 58
   nonincreasing function g, 51
   pseudomodular w, 50
Lipschitz constant, 99

M
mapping
   absolutely continuous, 109
   bounded, 79
   of bounded Φ-variation, 111
   of bounded φ-variation, 95
   of bounded variation, 104
   of Dirichlet-type, 85
   regulated, 82
metric, 1
   discrete, 9
   extended, 1, 9
   Hausdorff, 126
   uniform, 79
metric modular, 4
metric modular space, 21
metric semigroup, 127
modular, 1, 4
   $(a, 0)$-modular, 14, 31
   φ-convex, 46
   s-convex, 47
   canonical, 8
   classical, 11
   classical convex, 11
   complex, 48
   convex, 5, 27
   generalized Orlicz, 13
   Hausdorff, 68
   infinite, 9
   over a convex cone Λ, 47
   step-like, 9
   strict, 5
modular closure, 77
modular entourage, 67
modular equality, 70
modular interior, 76
multifunction, 104

N
norm, 11, 30
norm-differentiable, 114

O
operator
   additive, 100
   Lipschitzian, 99
   superposition, 100
oscillation, 79
   joint, 80

P
partition, 93
pointwise precompact, 87
pseudometric, 1
   extended, 1, 9
pseudomodular, 5
   φ-convex, 46
   p-homogeneous, 39
   Hausdorff, 16, 60

R
regularization, 7
right inverse of
   Hausdorff pseudomodular, 60
   nondecreasing function φ, 57
   nonincreasing function g, 51
   pseudomodular w, 50

S
selection, 104
set
   closed, 68
   compact, 104, 126
   modular closed, 76
   modular open, 73
   open, 67
   power, 15
space
   $\ell_\infty(x^\circ)$, 34
   ℓ_p, 32
   $c(x^\circ)$, 34
   Banach, vii
   metric, 1

space (*cont.*)
    metric linear, vii
    metric modular, 4
    modular, 4, 19, 21, 30, 48
    normed, 13
    reflexive Banach, 114
structural theorem, 113

T
topology
    antidiscrete, 74
    discrete, 74
    finest among, 75
    metric, 67
    modular, 73
triangle inequality, 1
    generalized, 37

V
variation, 104
variational core, 93, 110
velocity, 2
velocity field, 2

Printed in the United States
By Bookmasters